How To Understand The True Cosmos

How To Understand
The True Cosmos

Dr. Sol Aisenberg

To order additional copies of this book, contact:
Xlibris Corporation
1-888-795-4274
www.Xlibris.com
Orders@Xlibris.com
122885

Topics Covered

New Gravity at Cosmic Distances:
Understand cosmic gravity,
Understand the cosmos,
Remove mysteries and speculations in the model of the universe,
Why dark matter is not needed,
Learn the true cause of redshift,
Cosmic redshifts are not due to the receding velocity of galaxies,
Cosmic redshifts do not prove the expanding universe,
The Hubble constant cannot show the age of the cosmos,
Cosmic gravity explains tired light,
Understand quasars,
Understand black holes,
Solve Olbers' paradox about the dark sky,
Remove speculative belief about the big bang,
Learn why inflation is not true and not needed,
The true cause of the cosmic microwave background (CMB),
Understand why dark energy is not needed,
Why the 1978 Nobel Prize for physics and the CMB should be reexamined,
Why the 2011 Nobel Prize for physics and Dark Energy should be reexamined,
Learn why the cosmos is apparently closed.

Dr. Sol Aisenberg

Acknowledgements

I am pleased to acknowledge the patience of my wife, Dr. Ruth B. Aisenberg, who was understanding and permitted me to take time from my family, and who accepted my obsession with science and technology.

This book is dedicated to my wife Ruth and my children, Anne, Mark, and Amy, as well as my grandchildren, Lilly and Serena.

I also acknowledge and appreciate the contributions of past generations of scientists who provided the results of their studies of the cosmos along with their important observations and theories and speculations. These provided part of the resources for my contributions.

Preface

The current Standard Model of the universe contains many theories, mysteries, and unproven speculations, and this suggests that the Standard Model is wrong and need reexamination.

I will use the many existing and validated observations of others to provide some essays here that should clarify and replace many existing errors and speculations. Fortunately, some people have already made many valuable observations so that new observations may not be needed.

If a theory cannot be confirmed by prior or subsequent observations, then the theory should be abandoned, replaced, or corrected. Remember, observations must take priority over theories and speculations.

This collection of essays about cosmic gravity follows my earlier self-published book, *The Misunderstood Universe* © 2009, which resulted after several decades of studying part-time the many mysteries and errors and speculations in the standard model of the universe. This first book is available from Amazon on the Internet and other Internet sites, as well as from public libraries and some bookstores. Some of the material in the former book has been amplified and included in this second book. Additional essays and insights have been added.

Because the essays were written at different times and are intended to be read independently, there may be portions that are redundant.

After several decades of consideration, I realized that there are two major causes of the errors in the understanding of the current dogma about the cosmos (universe), both related to the use of Newton's universal law of gravity and Newton's gravitational constant G for observations at cosmic separations. These observations at cosmic distances are outside of the solar system and the observations used by Newton in formulating his law of gravity.

A result of using Newton's law of gravity and the gravitational constant at cosmic distances is the mystery of dark matter and also the mystery of dark energy, both of which can be explained by my concept of cosmic gravity. Serious errors related to redshift and the beliefs in its associated unproven velocity interpretation are also disclosed.

One cause of the errors is the assumption by many others that Newton's universal law of gravity and his gravitational constant Gn are also valid at cosmic distances, although only based upon observations of planets in our solar system. This assumption will be shown to be wrong in this collection of essays, and the mystery of dark matter will be removed.

The second and more serious cause of the many more errors about the cosmos are due to the improper use of additional observations by Edwin Hubble who, in the early 1900s, used a more powerful telescope to study the fixed stars in the sky and determined that they were actually collections of galaxies located at vast distances and consisted of hundreds of thousands of millions of stars.

He also measured the redshifts of the remote galaxies and determined that the redshifts were a linear function of their distance from the observer. Unfortunately, he speculated that the redshifts were due to the Doppler effect and that the galaxies were receding. He later expressed doubt about the Doppler-effect interpretation. However, others accepted this interpretation as fact, and this resulted in many errors in the understanding of the cosmos.

This led to the dogma that the galaxies were receding (speculation) and that the universe was expanding (speculation), and it supported the belief (speculation) about the big bang.

Note that there are no direct observations of the receding velocities.

The serious consequences of the speculative interpretation of the redshift include the need to reexamine the true meaning of the observed redshifts.

Theories and speculations are based upon past observations and must be confirmed by additional observations. When not confirmed, they must be replaced or modified.

While I am not trained as a cosmologist, my experience as a scientist is very broad, and it includes earning a PhD in physics and math from MIT and serving as a part-time staff member in the physics department of MIT and the MIT Research Laboratory of Electronics (RLE).

I was also elected as a member of Phi Beta Kappa, Sigma Xi (science), and Pi Mu Epsilon (math).

When I had the time to spare, I held part-time appointments as a lecturer at Harvard Medical School and as a visiting research professor at Boston University.

Most of my work was in industry, where I was involved in plasma technology, space technology, defense technology, and medical technology including diabetes, the artificial heart, and medical instrumentation. I was also the inventor of diamond-like carbon (DLC)

Details about my patents and reports, presentations, and publications are available in an appendix in this collection of essays, and larger lists are provided in my earlier book, "The Misunderstood Universe".

Introduction

I have been asked several times if my analysis of the speculations in the model of the cosmos (universe) has received peer review. As a longtime member of the American Physical Society (APS) and a reviewer for the Society of many papers submitted by others, I wanted to share my analysis with other physicists.

Thus years ago, as a scientist, I submitted several papers about my analysis of the cosmos to the APS for review and publication. They were returned (apparently without reviewer comments) because they were accused of being just speculations. Actually, they did not include speculations but were based upon many cosmic observations reported by others and were intended to answer the many speculations by others.

This is ironic because the standard model of the cosmos (universe) itself is full of speculation by others.

Therefore, in this collection of essays, I will take extra effort to label and identify statements and theories that are based upon observations and will label and identify other statements and theories of others that are only based upon speculations and that are without supporting observations and should be abandoned.

Apparently, it is difficult to overcome existing dogma. This is shown by the treatment by the astrophysics establishment to the brilliant Fritz Zwicky who, after his many contributions, was denied access to telescopes. Also badly treated was the very capable Halton Arp, who was denied access to telescopes in the USA and has to go overseas for access. In my essays, I will identify the truth and value of their contributions.

Copernicus also had problems with the dogma of his period. Galileo had even more problems and even physical threats from the later dogma of his time.

Because I feel that it is important, I wrote and self-published a book, *The Misunderstood Universe* © 2009, that presented my analysis at that stage.

This current additional book of self-published essays describes the role of gravity at cosmic distances for the understanding of the cosmos and adds some of my other insights. I hope to have future editions published by established publishers of scientific books for wider distribution.

These two initial self-published books may be of interest and value to future generations.

If I have time, I will submit several papers again to various international publications, but I will depend upon my published books and the postings on the Internet and maybe by recognized book publishers to preserve the solutions for future generations. *I will then return to my projects* for patenting and licensing my inventions. Planned is an invention that almost doubles the miles per gallon of cars and diesel engines. Also planned is an invention that can control cancer and drug resistant infections, without the need for drugs.

After years of working in science and technology as an applied and experimental physicist in many disciplines, and involved mostly in industry and on government programs, I gained experience in a large number of fields. Also involved in part-time academic activities with staff appointments at MIT, as a lecturer at Harvard Medical School, and as a visiting research professor in biotechnology at Boston University, I thus gained experience and knowledge including publication, reports and patents (over twenty-five US patents issued) in many fields, and could thus combine multiple disciplines in solving problems without requiring help from a multidisciplinary team.

Because of my curiosity as a scientist, I spent the last several decades in part-time study of the mysteries in understanding the cosmos. Many versions of my conclusions were presented earlier on the Internet.

To see the earlier web postings and other related information, you can easily do a Google search on Sol Aisenberg. Also try to see a copy of my first book on the subject, *The Misunderstood Universe* © 2009, 263 pages) by Sol Aisenberg. A copy is available from your library or bookstore, and it is available from Amazon as a softcover, hardcover, or as an e-book.

After this next collection of new essays is published, much of the first book may be made available for free download from the Internet, although you may find the cost of printing the 263 pages prohibitive. You will then have permission to freely share portions from this first collection of essays (unchanged and with proper attribution) with your friends and colleagues. They are also encouraged to freely share these portions of the first book.

Aisenberg, Sol. *The Misunderstood Universe.* © 2009 (263 pages)

This is an early collection of essays on the subject. Copies are available from Amazon and bookstores, or by loan from your library.

About the Author

With respect to his credentials for contributing to the understanding of the cosmos, the author is a scientist and physicist who earned a PhD in physics from Massachusetts Institure of Technology, (MIT) after graduating with a BS cum laude in physics from Brooklyn College in New York City.

Then at the graduate school in MIT, he held part-time appointments as a staff member in the physics department and in the Research Laboratory of Electronics (RLE) labs of MIT.

Dr. Aisenberg, when he had time, was also a part-time lecturer at Harvard Medical School and a visiting research professor in the bioengineering department of Boston University.

He was elected to the Phi Beta Kappa, Sigma Xi (science), and Pi Mu Epsilon (math) honor societies.

He was a senior scientist at the Raytheon research division. He then was the physics department manager at Space Sciences Inc. At the conglomerate Whittaker Corporation, he was the physics department manager, general manager, then division president and principal investigator in its space sciences division. Then at the conglomerate Gulf+Western, he was division president and principal investigator of the applied science division of the conglomerate Gulf+Western.

Dr. Aisenberg is a generalist, working and poblishing in many fields and disciplines including ultra-high vacuums, plasma physics, instrumentation, optics, solid state physics, semiconductors, thin films, space propulsion, electromagnetic theory and practice, energy conversion, lasers, diamond-like carbon (DLC), paramagnetic resonance, microwave systems, diabetes and glucose sensors, oxygen sensors, pulse rate monitors, lead poisoning detection, blood pressure instrumentation, biocompatible material, eye view monitors, dyslexia, medical devices, diabetes instrumentation, biocompatible materials, and the artificial heart program and other medical programs. This work resulted in many of his contract reports, plus publications, presentations, and patents.

After forming his own consulting company, the International Technology Group and its Data Associates division, he consulted for clients in Sweden, England, and in the United States, as well as for patent firms, including ones in Washington, New York, and Massachusetts. Aisenberg was a senior advisor on intellectual property for Thompson Financial and the Depository Trust Company.

Aisenberg has over 134 publications, presentations, and reports plus a number of invited talks. He has over twenty-five US patents issued, and others pending. He has obtained seven IR-100 awards for important new products and an award for research in helium-neon lasers. He was a reviewer for a number of technical journals and has reviewed proposals for the National Institute of Health and for the National Science Foundation.

Contents

Appendices

Essay 1

Gravity:
misunderstandings and the consequences

When Newton's law of gravity is applied to observations for galaxies and stars at large distances in the cosmos, it results in a large number of mysteries and errors in the model of the cosmos because of speculative interpretations of the new observations. In these essays, we will identify these errors and provide new interpretations based upon the observations without speculations.

The search for dark matter can end because it is based upon the failure of Newton's gravitational constant and his universal law of gravity to explain new observations of galaxies and stars at vast distances in the cosmos. When Newton developed his law of gravity, he only had available observations on Earth and for the observed motion of planets in our solar system. If, while he was alive, he had available the observations for cosmic distances, he could have arrived at a gravitational constant that increases linearly with distance r in the form $G=Gn+A^*r$, and that adds a gravitational force that is an inverse r force in addition to his inverse square component. This also extends the range of gravity attraction for cosmic distances and can speed up the formation of cosmic structures such as stars, galaxies, black holes, and webs of stars.

The corresponding forces for the new cosmic gravity constant including the inverse square distance dependence for smaller distances is F=M*G/r*r or F=M*Gn/r*r + M*A/r.

The term M*A/r shows that the additional cosmic force decreases much slower than the inverse square r dependence and has a longer range cosmic effect.

Even more important, the fruitless search for dark matter is no longer needed, and all the documents invoking dark matter should be rewritten. The energy and time and funding of scientists and engineers searching for dark matter should now be used for more serious matters.

A more serious mistake involving gravity is related to the consequences of the speculation by Hubble that the observed redshifts for remote galaxies were due to the Doppler effect. This resulted in the supposed support for the speculative belief in the receding velocity of galaxies and the expansion of the universe, although there are no known direct observations of the receding velocity of galaxies or of the expansion of the universe. Other consequences include the speculative age of the universe and the speculative interpretation of the cosmic microwave background (CMB). These will be discussed in other essays.

It will be shown that gravity can cause a redshift by extracting energy from the travelling photons. After my analysis of redshift and tired light, I found that the brilliant Fritz Zwicky had earlier suggested the role of gravity in the redshift. My analysis also considers the CMB and its temperature as the same as that of gas and dust in interstellar space.

One example of the contribution of gravity to redshift is the gravitational redshift observed for photons exiting from the sun.

The following is a more detailed discussion of the situation:

There are four fundamental forces in the cosmos. Gravity is the one that is apparent to all because it is experienced directly when a person walks or stands or climbs or falls. The other forces (the electromotive force, the strong nuclear force, and the weak nuclear force) can be sensed only by special means.

My analysis has shown that an incomplete understanding of the role of gravity has resulted in many mysteries and errors in the standard model of the universe (cosmos).

The genius Isaac Newton, using observations of the motion of planets in our solar system, developed his universal law of gravity which explained the observations of the motion of planets and the effects of gravity on our planet, the Earth. This law involved Newton's gravitational constant and an attraction that varied as the inverse square of the separation between two masses. This law worked for distances within our galaxy, the Milky Way.

However, when more precise technology became available in the last century, it was found that there were tiny deviations in our solar system when NASA space probes Pioneer 10 and 11 were sent into the solar system and each demonstrated a Pioneer anomaly where very precise measurements showed tiny deviations from their orbits as expected from Newton's law. Their observed locations indicated a tiny force toward the sun and suggested an additional force about eight orders of magnitude less than Newton's gravitational constant. Other explanations were investigated, such as gas leakage, to provide the extra central force for both probes.

This Pioneer anomaly suggests that the universal law of gravity needs reexamination and replacement or modification. This will be explained in other essays.

When Edwin Hubble had access to more advanced telescopes in the early 1900s, he found that the very remote fixed stars in the cosmos were actually galaxies, each consisting of a vast number of stars. The fixed stars (galaxies) were fixed in appearance because they were at cosmic distances.

This opened up the opportunity to investigate vast numbers of galaxies at far distances from our own galaxy and solar system.

The very capable Fritz Zwicky then carefully observed the motion and separations of groups of galaxies and found that the observations could not be explained using Newton's universal law of gravity unless vast amounts of invisible mass also existed. This was an early indication of the supposed need for dark matter. It also may have shown a need for reexamination of Newton's law of gravity for use in the cosmos.

Decades later, the capable Vera Rubin observed the motion of stars in spiral galaxies and determined that the stars were revolving at a constant velocity at an increasing distance from the center rather than with decreasing velocity with distance as required by Newton's law of gravity. She then suggested a halo of dark matter as an explanation for these observations. Actually, the missing mass and associated gravity should have been suggested for placement in the regions where constant velocities were observed rather than as a halo. Her observations of the rotation of spiral galaxies were important in developing the insight into cosmic gravity.

A more serious consequence of the speculation by Hubble and others that redshift was due to the Doppler effect was the resulting belief in dark energy, which involved the speculation that the supposed recession velocities of the remote galaxies and supernovae type Ia had accelerated, thus requiring dark energy.

The speculative belief in the receding velocity resulted in the speculation of the supposed age of the universe, the expanding universe, the big bang, and other beliefs that should be reexamined.

Essay 2

There is no need for dark matter

The genius Newton introduced his theory of universal gravity based upon available observations reported for the planets in our solar system. If Newton also had knowledge of the observations that were available much later for galaxies at cosmic distances he probably could have extended his universal theory of gravity to also include them in his universal law of gravity to make it a true universal law of gravity that can also explain the observations of the universe outside our solar system as well as in our solar system.

Long after Newton's death, in the early 1900s, the availability of more advanced telescopes permitted Edwin Hubble to observe the very distant fixed stars in the cosmos and find that they were actually collections of galaxies, each with vast numbers of stars.

Observations by the remarkable Fritz Zwicky of the motion of galaxies in remote groups of galaxies could not be explained using Newton's law of universal gravity, which was accepted as truly universal by most scientists. His careful observations showed the apparent need for a large amount of invisible matter to supply the necessary gravity for explaining these observations using Newton's law of gravity (for the solar system).

Decades later, Vera Rubin made detailed observations of the rotation of stars in spiral galaxies and found (and suggested) that the observations (confirmed by others) could be explained using Newton's law of gravity only if large amounts of invisible matter were present. A halo of dark matter around the galaxy was introduced by her to attempt to explain the surprising observations.

Actually, the supposed dark matter should have been proposed to be in the region of the spiral galaxy where the velocities were observed to be constant, not as halo of dark matter where its effect would not be required.

A modification of Newton's law of gravity (MOND) was introduced by Martin Milgrom in an attempt to explain the cosmic observations, but it involved acceleration and had limitations.

My analysis of the observations by Vera Rubin of the observed motion of stars in spiral galaxies considered the equilibrium between the gravitational attraction to the central mass M described by M^*G/r^*r and the centrifugal force of a star rotating with a velocity v at a radial distance r, described by v^*v/r gives the simple equilibrium equation $M^*G=v^*v^*r$, which is based upon standard physics.

For the observed case of constant velocity in spiral galaxies, the result is that the product M^*G is a linear function of distance r at cosmic distances. The usual belief is that the linear dependence is associated with the mass M leading to the unproven, speculative introduction of invisible dark matter and a longtime unsuccessful search.

I suggest a theory of cosmic gravity or additional gravity (TAG) where $G= Gnewton+A^*r$, which reduces to Newton's law of gravity for small r and which is valid in our solar system and also reduces to A^*r for cosmic distances where the inverse square dependence becomes small and negligible.

To determine the approximate value of A, look at the data from plot of the velocity vs. radius for the spiral galaxies NGC2403 and NGC3198l. The flat portion of the velocity curve starts at a transition radius Rt of about 2.7 kpc (kilo parsec) where the cosmic contribution to gravity starts to become greater than the gravity described by Newton. At the distance Rt, Newton's gravitational constant Gn is about equal to A^*Rt.

Thus $A = Gn / Rt$ and based upon the value of Newton's gravitational constant Gn and Rt, can be computed to be *about* $A=1.18 \times 10^{-14}/sec^*sec$.

I suggest that the linear dependence on distance is really associated with gravity G which itself is already invisible and is only detectable by its influence on visible objects. There have been decades of expensive searches for dark matter by excellent engineers and cosmologists including the very expensive Large Hadron Collider at CERN searching for evidence of the Higgs Bosons related to an explanation for dark matter. I believe that because of cosmic gravity, the dark matter mystery is solved.

Thus, my theory of cosmic gravity (or additional gravity, TAG) where G=G(n) + A*r with the constant A as a multiplier for distance r can explain the observations at cosmic distances and for small distances r and will also reduce to Newton's law in our solar system. The force F attracting masses will be the usual F=G(n)/r*r plus a force A/r that is still attractive but decreases slower with distance and thus has a longer gravitational range and significant effect in the cosmos.

When powerful computers used by others tried to model the structure of the cosmos using Newton's law of gravity in an attempt to duplicate the observed configuration of the universe, the result was unstable unless dark matter was included. The longer range gravitational component when used in computer modeling of the cosmos's structure may be a better alternative than the dark matter used in the successful computer modeling.

When the supposed dark matter is introduced in the computer modeling the result is more satisfactory. Introducing the additional gravity term A*r rather than the supposed dark matter should also provide satisfactory stable modeling of the structure of the cosmos including the galaxies and strings and walls of galaxies.

Also when the distance r is reduced by many orders of magnitude to correspond to the distances of the NASA Pioneer space probes 10 and 11 in our solar system, then this smaller additional gravity can explain and predict the Pioneer anomaly, which is the observed (orders of magnitude smaller) additional attraction of the space probes toward the central sun.

This additional gravity would also agree with the observed gravitational cosmic lenses as suggested by Einstein and without needing dark matter.

Based upon the above analysis, I suggest that the mystery of dark matter should be clarified in other books and articles so that others will not waste time money and effort on this apparent mystery.

Essay 3

Dark energy is wrong

Based upon publicly available observational data, I will attempt to provide solutions to mysteries in the model of the universe. This will show why the belief in dark energy should be reexamined. In true science, when theory involves unproven speculations or does not agree with reliable data, the theory should be modified.

Observations take priority over theories and speculation.

The concept of dark energy is based upon the belief that Edwin Hubble's observations in the early 1900s of redshift as a linear function of distance can be used to determine and demonstrate the receding velocity of galaxies and the expansion of the universe. He called redshift an apparent Doppler effect but later expressed doubt about this speculative interpretation.

Apparently, there were no reports of direct observations of the receding velocity of galaxies.

There are current beliefs and speculations that the redshift showed that the galaxies were receding and that the universe is expanding. This is a major cause of errors in the model of the universe.

There are other possible causes for a redshift such as the observed gravitational redshift of light from the sun and the loss of photon energy through gravitational interaction of photons in tugging and moving low-mass gas and dust while traveling large distances through

29

interstellar space. The contribution of gravitational drag by the gas and dust on traveling photons is described in another essay.

Collisions or absorption of photons would have resulted in fuzzy images of the source stars and is not an acceptable explanation.

The belief that the galaxies were receding and the universe is expanding even fooled Albert Einstein, who introduced his cosmological constant in his equations because of his belief (or wish) for a constant universe. With the demonstration by Hubble of the apparently receding galaxies, Einstein then removed his cosmological constant and called it his greatest blunder.

I believe that the question of an open, flat, or closed universe is still not solved, although the observed progressive clumping of gas and dust into dense structures such as stars, galaxies, strings, and walls, may be the start of the universe closing, which can take much longer time than the lifetime of an observer or the duration of civilizations.

Later, in about 2011, a difference was found between the observed distances of very faint remote standard candles, supernovae type Ia, and their distances based upon their observed redshifts. Note, however, that for very remote distances, the distances determined for the standard candles type Ia had been used to calibrate the distances for remote redshifts.

It was speculated that these galaxies had accelerated and needed dark energy. I believe that this may be wrong because it depends upon an assumption of receding velocity. This assumption apparently is not validated by observations of velocity but only an assumption of the Doppler effect as a cause for the observed redshift.

This discussion predicts that for very remote galaxies, there will be differences between the distances of supposedly receding galaxies as determined from observed magnitudes and the distances determined from their observed redshifts.

The difference between the distances resulted in a claim that the remote galaxies had accelerated, thus requiring dark energy for the acceleration of the supposedly receding galaxies.

I believe that the basic assumption of receding velocity that is the foundation for the need for dark energy may not be accurate.

Much of the background of dark energy is described in interesting detail by Professor Robert P. Kushner in his book *Extravagant Universe*, which is available from Amazon and their out of print resources.

Kushner, Robert P. 2002. *Extravagant Universe.* New Jersey: Princeton University.

Essay 4

Redshift errors

I believe that, starting in the early 1900s, the work on the model of the universe lost its way. The introduction of the mysteries of dark energy and dark matter should have alerted people that the model of the universe needed correction. Observations must take priority over theories and speculations if there is no agreement. I also believe that the use of singularities in theories is also a sign of a lack of associated knowledge.

Many of the errors in the model of the universe are based upon the belief that Edwin Hubble's observations in the early 1900s of redshift as a linear function of distance can be used to determine and demonstrate the receding velocity of galaxies and the expansion of the universe. He called redshift an apparent Doppler effect but later expressed doubt about this interpretation.

Actually, there were no reports of direct observations of the receding velocity of galaxies.

There are other causes for redshift such as the gravitational redshift of light from the sun and the loss of photon energy by interaction of photons in moving low-mass gas and dust by gravitational attraction, which extracts energy without needing actual collisions while traveling large distances to the observer through interstellar space. Note that in this collection of essays it was shown that observations for cosmic distances, there is an additional gravitational force that decreases inversely with separation r, rather than the inverse square dependence with distance. Thus the gravitational interaction of traveling photons with interstellar gas and dust will be larger because more gas and dust further away will gravitationally interact with the photons and increase the redshift. As a result of the belief that the redshift was due to the receding velocity of observed galaxies, I believe that there are now unproven and seriously wrong features in the model of the universe, including the following:

(a) Receding galaxies
(b) The expanding universe
(c) The big bang
(d) Violation of physical laws in the first fraction of a second, in the big bang, such as the expansion velocity greater than the velocity of light c
(e) Violation of physical laws by initial expansion against the massive gravitational attraction of the supposed massive and very tiny and compact energy-mass of the big bang
(f) Need for inflation to explain the universe uniformity beyond the event horizons
(g) Cosmic microwave background (CMB) as remnant of the supposed very high temperature of the big bang
(h) Age of the universe based on the use of the Hubble constant in the velocity form resulting in the form of the Hubble constant as inverse time. Inverting the Hubble constant when defined in terms of velocity gives it the dimensions of time and the apparent age of the universe. I believe the supposed age of the universe is wrong because of speculation about an apparently unproven Doppler effect and the resulting velocity assumption, but without any direct observation of receding velocity.

Edwin Hubble—in his book *The Realm of the Nebula* (New Haven: Yale University Press, [1936] 1982), page 122—indicates reservation about the velocity interpretation of the redshift by including: "judgment may be suspended until it is known by observation whether or not red-shifts do actually represent motion."

Essay 5

Showing that gravity contributes to the redshift

The belief in the redshift observations by Hubble is wrong and has resulted in many errors in the supposed understanding of the cosmos.

We will show that redshift does not represent the receding velocity of galaxies as was suggested by Hubble as being due to the Doppler effect. This questions the speculations that the universe is expanding and that there was a big bang.

We will also show that the redshift cannot even be used to determine distance in some cases and that cosmic gravity takes an important role in the redshift.

Consider the case of the observed redshift of light from our sun received by observers on Earth. If the observed redshift represents receding velocity, then the wrong conclusion is that the sun is receding from Earth, contrary to experience. (We would slowly freeze if this is the case.)

One could counter that redshift is due to the speculation that the universe and space-time is expanding, thus increasing wavelengths and causing redshifts. This is an example of circular reasoning: the speculation that the universe is increasing is based upon the belief that redshift shows that the galaxies are receding and that the universe is expanding, followed by the claim that the expanding universe caused the wavelength to increase, resulting in the redshift. Thus, this example of circular reasoning fails to counter the objection involving the redshift from the sun.

With respect to the use of the redshift to determine the distance to the source of the light reaching the observer, if the source of the light also contains or is influenced by a nearby massive dark hole or other large collection of matter, the large gravitational effect will also contribute to the redshift in addition to other causes of redshift similar to the gravitational redshift from our sun.

One example involves a galaxy that contains a massive black hole, and that radiates large amounts of radio-frequency energy and is called a quasar. As described in another essay, the black hole will contain matter that is at high temperatures, such as stars where the energy distribution is in thermal equilibrium such as a Maxwell-Boltzmann distribution. Photons emerging from the gravitational well will come from the high-energy tail of the distribution and will lose most of their energy and arrive in the radio-frequency range, thus appearing as from a radio-frequency galaxy and called a quasar.

A redshift from a star or galaxy thus cannot always be used to accurately to determine its distance.

In fact, because of the strong gravitational effect on the redshift signal arriving, possibly from other sources in the quasar, the distance to the quasar will be determined to be much larger than its real value. When this false distance and the light from this quasar are used to determine its energy output, the result will be that the energy output from the quasar appears (speculation) to be unusually large and this is incorrect.

Another demonstration that the distance to the quasar as determined from the observed redshift is incorrectly large is when the observed angular (sideways proper motion) velocity of the quasar is combined with the apparent distance to find sideways velocity; it is computed (speculation) to be greater than the limit of the velocity of light. Thus, redshift cannot accurately be used for quasars or galaxies containing large mass.

Our analysis indicates that the redshift of photons travelling large distances through interstellar space will have their energy depleted through gravitational interactions with low-mass gas and dust in their passage through interstellar space. Energy transfer between two entities is more effective when the masses are closer to equal. The loss of energy with distance will result in redshifts that increase with distance. The gravitational coupling will be more effective for photons that travel closer to interstellar gas and dust, and some of the photons travelling closer to the gas and dust will come into thermal equilibrium with the temperature of interstellar gas and dust, which was determined by other means as being about 5 K. (see list of references and reading material) This will explain and predict the observed cosmic microwave background (CMB). The association of the CMB with the remains of the speculated big bang is wrong and not needed.

In 1978, a Nobel Prize in Physics was awarded to Penzias and Wilson for their work in detecting the cosmic microwave background. Subsequently, use by others of the observed CMB to support the speculation that it was the result of the big bang thus led to additional errors in the model of the cosmos.

Conclusions: Redshift is not significantly due to velocity, and gravity contributes to redshifts; thus, redshifts cannot always be used to determine distances, or receding velocity.

An example is the observed gravitational redshift of spectral lines from the sun. Fortunately for those on Earth there are other contributions to the redshift for larger distances, and the sun is not receding.

Essay 6

Standard candles: supernovae type Ia (SN Ia)

The use of supernovae type Ia as standard candles to determine the distance of remote galaxies is important and valuable, but there is a question of why the observed light output of different SN Ia light sources will have different rates of decline of light output. There is an observed correlation between the peak brightness of SN Ia and the rate of approach to peak brightness as well as the rate of decay of light after maximum brightness. Brighter supernovae brighten slower and decline more slowly. In contrast, fainter supernova brighten faster and decline faster.

I suggest that the difference between the brighter SN Ia and the dimmer SN Ia is that the dimmer types of SN reach the critical mass factor of 1.4 identified by Chandrasekhar by slower collection of additional mass, while the brighter and larger SN reach critical mass much quicker through a larger addition of mass by collision with an asteroid or other massive object(s), and thus will quickly have more mass for the SN to use and be brighter. For larger sizes it takes more time for photons to travel and explains slower rise and fall times for larger and brighter SN.

I suggest that the less luminous faster SN Ia should be used as standard candles even though they are harder to detect and use. This avoids the potential complications of going super novae by collisions of different masses.

Consideration of the physical processes involved leads to the suggestion that the brighter supernovae with masses much in excess of the critical mass are also larger and that it takes more time for the photons to exit the larger supernovae because of longer paths and scattering collisions resulting in a slower decay. This is similar to the long time for photons to exit from the center of our sun. Also, because longer wavelength photons are scattered less on the way out, the time decay of infrared photons will be less than for red colors, which in turn is less than for blue colors.

The more luminous SN Ia take longer (about 70 days) to reach peak brightness and longer (about 250 days) to decay to about half peak brightness. This is in contrast to the less luminous SN Ia, which takes about 10 days to reach its peak brightness and only about 20 days to decay to about half its peak brightness. (*Scientific American*, June 2012, pp. 45-49)

Essay 7

Gravitational drag on photons causing redshift as a linear function of distance

Many competent scientists believe in the current dogma that the redshift observed by Edwin Hubble is due to the Doppler effect and that the galaxies are receding and that the universe (cosmos) is expanding after the supposed big bang. Hubble later expressed doubt about the cause of the redshift as due to the Doppler effect.

I will provide an explanation of the linear dependence of the redshift on distance as being due to gravitational drag on travelling photons by gas and dust in interstellar space.

This will also explain why the observed cosmic microwave background (CMB) exists and is observed to be in thermal equilibrium and at a temperature of about 2.75 K.

The Swiss astronomer Fritz Zwicky, whom I respect as an unappreciated true genius, had suggested the effect of gravity on photons earlier in 1929, and on the possibilities of a gravitational drag on light.

He was reported to be very grumpy, and this probably was because many of his valuable observations and contributions had been ignored and he was later denied access to telescopes.

After some time, redshift was called tired light, but this labeling did not help in the understanding of tired light. Just labeling something does not explain it.

In this essay, I will present an explanation for redshift based upon known physical processes without any speculations.

I will suggest the concept of gravitational friction by gas and dust in the interstellar voids where photons lose photon energy by gravitational interactions with gas molecules and dust on the way to the observer without requiring collisions.

After the photons continue, the interstellar gas and dust left behind will have been moved by non-collisional gravitational action and the energy required for moving the gas and dust will have been extracted from the photons.

Some of the photons that are sufficiently physically close to interstellar gas atoms or molecules will come into thermal equilibrium with them and at the interstellar temperature of about 2.75 K of the gas while photons that are not close enough to the interstellar gas to equilibrate will continue to longer wavelengths such as in the radio-frequency range. The gravitational range of interaction by cosmic gravity is greater than predicted by Newton's universal law of gravity because, as is shown in this collection of essays, the cosmic gravity force decreases as $1/r$ and is slower than the usual inverse square of distance. The amount of interstellar gas is reported to be greater than for the small dust. The transfer of momentum is more effective the closer the bodies are in mass, and thus, gas and photon interactions are more pronounced.

Thus photons from stars in remote galaxies traveling through interstellar space towards the observer will interact with the gas and dust in the interstellar space through gravitational interaction as they pass them, thus transferring and losing some photon momentum and energy as they pass by without needing collisions that would result in fuzzy images.

The suggestion of photon absorption and then later reemission would also produce fuzzy images of the stars. This is not observed and thus is not a valid possibility

In two-body interactions, these interactions are strongest when the masses of the photons and gas and dust are close. The average direction of the photon path is not changed because of the averaging of many small effects.

As the photons interact with interstellar gas and dust they lose energy and momentum proportional to the numbers of interactions and the path length (distance). Because the velocity of light is constant at c, they can lose energy only by reduction of photon frequency f (through

$E = h^*f$), corresponding to a loss of energy and an increase in wavelength (redshift).

This predicts and explains why redshift is a linear function of distance.

There now is no need to invoke the Doppler effect and the receding velocity of galaxies as a cause of the redshift. Actually, there are no reports of the direct observations of receding galaxies.

The speculation and belief that the redshift is due to the Doppler effect and is speculated to be related to velocity is wrong and has caused serious errors by cosmologists in trying to explain the observations, including the expanding universe, the big bang, and the age of the universe.

The resulting speculation about the receding galaxies is wrong and questions the validity of the recent Nobel Physics prize (in 2011) involving the accelerating expansion of supernovae type Ia and the need for dark energy. Note that if there are no observations of receding galaxies or of receding supernovas type Ia then there is no observational evidence for the speculated expansion.

Zwicky, Fritz. 1929. "Redshift of Spectral Line." *Proceedings of the Natural Academy of Sciences of the United States of America* 15:773-9.

Zwicky, Fritz. 1929. "On the Red Shift of spectral lines through interstellar space." *Proceedings of the Natural Academy of Sciences of the United States of America* 15:773-779.

Zwicky, Fritz. December 28, 1929. "On the Possibilities of a Gravitational Drag of Light" *Physical Review Letters* 34.

Essay 8

The age of the universe
based upon the redshift is wrong

The age of the universe is computed from the Hubble constant, but it is not correct because it is based upon the speculation that redshift is due to the Doppler effect and velocity and thus shows that the galaxies are receding. The Hubble constant using dimensional analysis has the dimensions of inverse time and is used to speculate on the age of the universe.

The usual analysis uses the reciprocal of the Hubble constant to give the age of the universe as about 13.7 billion years.

This is wrong because there is no evidence that redshift is related to velocity.

One consequence is that the resulting value of the event horizon (equal to the velocity of light c times the age of the cosmos) is wrong. However, the size of the cosmos will still be much larger than any corrected event horizon.

Conclusions involving the age of the universe will have to be reexamined.

It has been reported that there are stars older than the supposed age of the universe, and their ages were determined by measurement of various radioactive by-products in these stars.

Essay 9

Olbers' paradox: why the sky is dark

Years ago, Heinrich Wilhelm Olbers introduced a paradox as to why the sky is dark at night in spite of the billions of stars in the sky. With the large number of stars in the sky, there was a question about why the sky is dark when our sun goes down because, eventually, the line of sight must see a star.

It was suggested that the light from more remote galaxies were blocked by closer galaxies in the line of sight, but this was refuted by the fact that the blocked light was replaced by light from the occluding galaxies.

It was suggested that light from very remote galaxies had not yet reached the observer because of the limit of the event horizon and the limit by the velocity c of light.

I have suggested that because of the redshift of light from remote galaxies, the wavelength of this light had shifted outside the visible range of the eyes of human observers and could only be seen by other detectors outside the human visible range such as infrared, microwave, and radio-frequency receivers.

Some of the photons arrive in the ultraviolet range above the range of human visibility, and some of the photons have traveled below the range of human visibility.

The optical range of human visibility is normally limited to the blue down to the red.

This explains and predicts why the sky is dark at night when viewed in the visible range: *most of the arriving photons are outside of the human visible range.*

Essay 10

The Pioneer anomaly removed

It was observed that the NASA space probes Pioneer 10 and Pioneer 11 demonstrated anomalous motion in the sense that longtime precision measurements of their position in space showed that they experienced an unexpected and unexplained but very small attraction toward the sun. This force on both probes was many orders of magnitude smaller than expected from Newton's law and gravitational constant.

The two spacecraft were launched in 1972 and 1973, and deviations from predicted accelerations were observed after they passed about twenty astronomical units (AU), where an AU is the distance from the Earth to the sun.

Both radio Doppler and ranging data provided precision information for the velocity and distance of the spacecraft. Multiple measurements over many years showed that a very small but unexplained force remains, corresponding to a central acceleration towards the sun of $8.74 * 10^{-10}$ m/s*s and for both spacecraft.

Over a period of one year, they are calculated to be about 400 km closer to the sun than the predicted distance based upon Newton's law and constant. It was reported, "Some extra tiny force—equivalent to a ten-billionth of the gravity at Earth's surface must be acting on the probes, slowing their outward motion." John D. Anderson and his team at the JPL made refined analyses which ruled out a number of possible explanations of this extra force.

Suggestions had been made that the tiny additional force was due to gas leakage. This is probably unlikely because it would require that leakage be in the correct direction for both vehicles, which however are changing their orientations in space.

Recently it was suggested that the problem was solved if one considers the radiation pressure due to thermal radiation from one outside surface (and for both vehicles) is related to a temperature difference between radiating surfaces facing the sun and larger radiation from radiating surfaces facing away from the sun. This suggested solution involved circulations of the temperature profile in the vehicles. It also required proper orientation of the radiating surfaces. Because of the factors cited here, I believe that this is wrongly suggested as the solution.

I will present here my explanation for the increased gravitational attraction. It is based upon the cosmic gravity described on other essays to remove the need for dark matter. Briefly, the observed flat velocity rotation curves for spiral galaxies reported by Vera Rubin are explained by properly interpreting the equilibrium equation balancing the gravitational attraction and the centrifugal force of rotation which results in $M*G=r*v*v$, where M is the central mass, G is the gravitational constant, and rr is the distance.

For the observed constant velocities v for r starting at an observed r value of 2.7 kparsecs, (kpc) this reduces to the product of M and G as a linear function of r. The usual interpretation is to assign the distance dependence to the mass, resulting in the need for dark matter and the long and unsuccessful search for dark matter. In part because dark matter has the same properties as gravity (not emitting light, non-reflective, not eclipsing light, and only apparent by its effect on visible light-reflecting planets and stars) I believe that the distance dependence should be assigned to the gravity term G and not to the mass M term, thus also removing the need for the expensive search for dark matter.

Thus Newton's gravitational constant Gn is now the cosmic gravitational constant, Gn, which is now

$$Gc=Gn+A*r$$

where A is a calibration multiplier for r, the distance/separation. This equation explains observations at cosmic distances and also reduces without change or splicing to Newton's formulation in our solar system.

It is different from the interesting MOND theory of M. Milgrom which involves acceleration and a modification of Newton's law.

This term A*r becomes apparent at a distance of about 2.7 kparsecs when it becomes larger than Newton's gravitational contribution. We can evaluate the size of the A*r term at the distance of one AU corresponding to the Earth distance to the sun.

One parsec is equal to 206,265 AU and 2.7 kparsecs is equal to 5.569155×10^8 AU. Thus an AU for the value of cosmic gravity, A*r, when reduced by the distance difference is $1/5.569 \times 10^8$ or 1.796×10^{-9}. The latest measured distance for the probes is about 20 AU, and when reduced by the product $20*20 = 400$ is $1.796 \times 10^{-9} / 400 = 4.49 \times 10^{-11}$.

The latest location of the probes is said to be about 20 AU. Compared to the value of the cosmic gravity st the Earth, and using the inverse square rule, the value of the cosmic gravity is about $20*200$ or a factor of 400 times weaker or about 5.569×10^8. This is close to the already-cited estimated 10 billionth factor of the gravity at Earth's surface, particularly in view of the many orders of magnitude involved in the comparison.

Thus this effect of cosmic gravity is many orders of magnitude smaller in our solar system and similar to the tiny anomalous gravitational effect on the Pioneer probes.

It is difficult to compute the velocity related to the cosmic gravity effect because the velocity depends upon the mass involved and the masses are not specified in the other analysis.

Because of the much larger mass of the planets and the lack of similar precision equipment on the planets, this cosmic gravity effect will not be apparent for the planets.

Essay 11

Cosmic microwave background (CMB)

Observations of the cosmic microwave background (CMB) coming from all directions in the cosmos have been described as the remnants of the supposed big bang that started the universe. In our essay on the errors in assigning the redshift to the Doppler effect and the receding galaxies, we question the belief (speculation) in the expanding universe and thus question the need for the big bang and the associated belief that the CMB is due to the big bang.

We can now suggest a way that the travelling photons can result in the cosmic microwave background without the big bang. As the photons travel long distances, they fall below the visible range of humans. They reach the infrared range, then the microwave range, and finally fall into the radio-frequency range. There are large expensive arrays of steerable radio-frequency dish antennas being deployed.

Some of the photons, however, that are in closer proximity to the interstellar gas and dust will come into thermal equilibrium with them at the temperature of the interstellar gas and dust. The amount of gas in the interstellar space is much larger than for the dust. Independent calculations by others of the temperature of interstellar gas in response to radiation from stars show that is expected to be about 2-5 K. See the reference below. Use a Google search for access to the referenced document.

The observed cosmic microwave background show that the photons arriving as microwaves are quite uniform and at a thermal equilibrium temperature of about 2.73 K, which is close to the expected temperature of the interstellar gas.

The photons that do not equilibrate with the temperature of the interstellar gas (because they do not travel close enough to the gas) will continue downward in energy and continue to lose energy with non-equilibrium gravitational drag or friction, down to the radio-frequency range of longer wavelengths.

There is a question about what happens when the energy approaches zero and the redshift wavelength approaches infinity. We will leave this question to others.

There was a Nobel Prize in Physics related to the cosmic microwave background (supposedly supporting the big bang) awarded in 1978 to Penzias and Wilson, and this should be reexamined although the observation of the cosmic microwave background itself probably deserves the prize, but the resulting interpretations probably are wrong.

Assis, A. K. T. and M. C. D. Neves. July 1995. "History of 2.7 K Temperature Prior to Penzias and Wilson." *Apeiron* 2:79-84

Describes the important independent determination of the temperature of interstellar space.

Essay 12

Black holes and the information paradox

Black holes are reported to be at the center of many galaxies. Because of the large amount of mass in the black hole and the very strong gravitational well, additional nearby masses, including stars, are sucked and trapped in the black hole, thus helping it grow.

Stephen Hawking and others have seriously considered the question of the information paradox and the black hole. Hawking worked on the black hole problem with Roger Penrose and Kip Thorn, and there were several disagreements.

Respected physics theory requires the preservation of information, and there is a question of recovering the information that may be trapped in the gravitational well.

Stars trapped in the potential well are spiraling down. Some have not yet reached the bottom of the well, and the energy barrier they see against photon escape is not yet large to prevent some photons from the Maxwell-Boltzmann energy distribution of the hot stars from overcoming the potential barrier.

Because of the Maxwell-Boltzmann energy distribution in stars, even stars at or near the bottom of the gravitational well will have energetic electrons and photons that have enough energy to overcome the gravitational potential barrier and escape, carrying information out. The high-energy escaping electrons and photons are continuously replaced by the equilibrium processes in the Maxwell-Boltzmann energy distribution so that the process of replacement of energetic photons will permit the continued release of photons and information.

The photons that overcome the gravitational energy barrier will have their energy diminished, but higher energy photons in the very high-energy tail of the energy distribution can escape with extra energy (shorter wavelength) and with lower remaining energy. This can continue as long as the trapped stars radiate in the black hole and the information continues to escape.

The information escapes as photons break free from the gravitational trap in the black hole. The photons, when they escape, come from the high-energy tail of the Maxwell-Boltzmann energy distribution and lose much of their energy because of the gravitational barrier. Most of them are low-energy photons in the radio-frequency range, although some of them come into thermal equilibrium with the gas and dust in interstellar space and add to the cosmic microwave background (CMB) at a temperature of about 2.75 K. Radio frequencies can contain much information.

This explains and predicts the CMB and its temperature and its Maxwell-Boltzmann energy distribution for the detected microwaves.

As the photons and energy escape from the black hole, this reduces the energy mass in the hole and reduces the gravitational attraction of the trap. If the rate of mass evaporation by escaping photons is larger than the rate of outside mass entering the black hole because of depletion of sufficiently nearby masses, then there is a net loss of mass and the black hole can eventually vanish with the possible remains of a neutron star.

The information paradox is explained by the transport of information in the electromagnetic form of escaping photons.

Stephen Hawking has worked on the information paradox of black holes for some time, and he is recognized as a genius. He provided various explanations and recently withdrew his suggestions at a meeting where he was reported to substitute the possibility of multiple universes. This shows that even the most brilliant scientists are reduced to speculations.

Essay 13

Quasar properties are wrong

Quasars are galaxies that are determined to be emitting much more energy than regular galaxies. This is based upon the output energy received from the quasar and calculated from the distance determined from its observed redshift. The output energy of the galaxy can be calculated from the output energy (luminosity) and the apparent distance determined from the observed redshift. This results in the calculated value of the energy output of the galaxy to be very large, which qualifies the galaxy to be termed a quasar.

However, because the quasar may also contain a massive black hole, the gravitational attraction of this black hole may increase the redshift and may make the distance of the galaxy appear to be much larger than the real value. As a result, the computed energy output of the galaxy is in error. In fact, the distances based upon observed redshifts for some galaxies may be wrong and too large if they also contain massive black holes.

There is another problem caused by distances determined for quasars with large black holes. When the transverse velocities of some quasars determined from the observed angular transverse motion in degrees per unit time are multiplied by the distance deduced from the redshift, the transverse velocity is sometimes calculated to be larger than the limit of c, which indicates an error probably due to the wrong distance.

References for these problems are provided in several books by Halton Arp, who apparently has had problems with the current dogma and acceptance of his observations. These books are listed in the references and in Appendix A for additional reading.

One interesting observation that he has made and described is that of two galaxies connected by a stream of stars from one galaxy to the other. This suggests that they are located near each other. However, one of the galaxies has a distance determined by redshift to be much larger than for the other observed (connected?) galaxy. As a result, apparently the establishment and reviewers have rejected Arp's important observation and limited his access to telescopes. This is a disturbing action.

I suggest that the problem with accepting his observations is due to the situation that the galaxy that is apparently farther away has a massive black hole, which adds to the redshift and makes it appear that the galaxies are far separated.

Another problem that Arp encountered is when he cited observations that showed periodic gaps in the observed redshifts and their apparent velocities when viewed in single directions. Reviewers of his contribution countered that when the views were made in many other directions, the periodicities were not observed. The reason for this is obvious if one considers that the periodic gaps may not have the same starting points for different directions. This is similar to the case of a sponge with holes that are present in slices of the sponge but cannot be seen when the slices are overlaid because of displacement of the holes. Actually, the gaps in the redshifts and assumed velocities is probably due to periodic gaps in the distribution of stars and galaxies in the cosmos, as can be seen from viewing the patterns of strings and walls in the cosmos. Because the redshifts depend upon the distance travelled by the photons, and there are gaps in the spacing of galaxies, this can explain and predict the observed gaps in the redshifts.

This is another example of the rejection of observations by skilled observers, rejected by recognized experts who are committed to the current dogma.

Interesting details are provided in the books referenced below and in the list of reading material in the Appendix.

Arp, Halton. 1998. *Seeing Red: Redshifts, Cosmology and Academic Science.* Montreal: Apeiron.

Also http://redshift.vif.com

Arp, Halton. 1987. *Quasars, Redshifts, and Controversies.* N.p: Interstellar Media.

Essay 14

The big bang dogma is wrong

Some of the major and serious consequences of the wrong use of the observed redshifts being interpreted as being due to the Doppler effect includes the incorrect speculation that the galaxies are receding, that the cosmos is expanding, and that there was a big bang.

The amount of personal energy, time, and federal money that has been devoted to work based upon belief in the big bang is distressing and may be a waste of limited funds. Such errors should be avoided in the future.

Another result of the apparent violation of basic laws of physics was the speculation about the very large expansion speeds of energy and entities spreading through the cosmos at the start of the big bang (as being larger than the limit of the velocity of light c). This apparent violation of basic physics was ignored.

Also a violation of a basic law of physics was the speculated rapid expansion against the massive gravity of the speculative and very compact large energy and its equivalent mass ($E = m \times c \times c$) at the start of the big bang, and this violation of basic physics and gravity was also was ignored.

Essay 15

Inflation explanation is not needed

The event horizon is defined as the distance that light or other effects can travel based upon the product of the velocity of light c and the age of the universe. Observations indicate that the extent of the universe/cosmos is many orders of magnitude larger than that of the event horizon. Yet other observations indicate that the different regions of the universe separated by the distances of many event horizons are apparently similar although they cannot communicate or interact, and are limited by the velocity of light and the age of the universe and the event horizon. The concept of inflation to explain the uniformity was introduced by Professor Alan Guth, now at MIT, and is described in his 1977 book.

The concept of the big bang is that the universe originated from a very tiny spot (singularity) and rapidly expanded in a very, very small fraction of a second. Thus there is a belief that the universe appeared in a speculated big bang in the form of vast energy (equal to vast matter by E=m*c*c) and expanded in very, very small fractions of a second to form the observed mass in the universe. The inflation argument basically is that the energy at the start of the creation and subsequent expansion was in initial equilibrium, and this equilibrium and uniformity was preserved during the expansion to give the observed eventual uniformity in the current universe.

This appears to violate some basic laws of physics. The supposed expansion of the initial energy is in conflict with the limit on the velocity of light c which also limits the speed of interaction. Also in violation is the apparent ability of inflation to require the vast amounts of mass/energy to overcome the effects of massive gravity associated with the initial vast mass/energy.

In order to believe in inflation to explain the observed uniformity of the universe and the supposed big bang, we will be required to abandon the belief in the limit of the velocity of light and the effect of gravity. It was suggested to me during a physics meeting that these laws of physics were not operative during the initial phases of the big bang and expansion. Use your own judgment.

If one subscribes to the concept of a big bang occurring in a small fraction of a second (without observational support) we can also speculate and introduce the concept of a sequence of multiple big bangs in a fraction of a second also witout observational support. This is an equally valid speculation, but one that I do not support.

There is another potential explanation for the uniformity of the universe within the limitations of the event horizon. (Note that the age of the universe is probably wrong because it depends upon the belief that the redshift involves velocities of the receding galaxies and the Doppler effect explanation.)

Consider the case of an event horizon region A, which mostly overlaps with an adjacent event horizon B, which in return overlaps with another event horizon. Then when region A mostly overlaps with region B, the result is that A and B are in equilibrium and the uniformity extends beyond the event horizon limit for each. This is also true for the regions B and C overlap. Then because A = B and B = C, we get the logical result that A = C. By covering the universe with a series of mostly overlapping event horizons, we get the explanation for the observed uniformity of the universe without requiring inflation or violation of two basic laws of physics.

However, there is still the problem of the speculation about the big bang without confirming observations, except speculations about the observed consequences such as detecting hydrogen, helium, and lithium. There can be other reasons for detecting hydrogen, helium, and lithium.

Guth, Alan. 1977. *The Inflationary Universe*. Cambridge, MA: Perseus Books.

Essay 16

The fate of the cosmos/universe: is it open, flat, or closed?

There is concern about the eventual fate of the cosmos/universe. We will provide arguments citing observations that show that the universe is closed.

I believe that speculation about the beginning and end of the universe (cosmos) is foolish because apparently, we do not understand the present universe (cosmos). However, we will continue here.

Open universes will continue an apparent expansion.

A flat universe will not expand or contract. This was proposed by Einstein who introduced his cosmological constant into his equations to support his belief (hope) for a stable universe.

A closed universe will collapse under gravitational forces to a very tiny universe sometimes identified as a singularity. I believe that suggesting a singularity is the same as saying that "I do not know".

The introduction of a singularity in a theory about the beginning of the universe shows that the theory is not really understood and needs further analysis.

The information about the cosmos and also about the cosmos millions and billions of years go (due to the time required for light to travel) shows that matter started to collect into stars and galaxies, black holes, and structures (such as strings, walls, and voids) many years ago.

This light received from the current and past cosmos shows that it is collapsing. Observations of the structure of the visible cosmos shows that much of the dust and gas are collected by gravity into stars, which are then collected in the form of galaxies, and then into cosmic structures in the form of cosmic strings and walls of galaxies.

Many of the galaxies indicate that they contain black holes, which are massive collections of matter and have strong gravitational effects.

Cosmic observations show that much matter has already collected in the form of strings of galaxies and walls of galaxies and large voids empty of matter.

This appears to be an indicator of the process of the cosmos collapsing and a sign of a closing cosmos.

All these observations show that the cosmos is collapsing into collections of visible matter which, in turn, will continue to collapse together because of the gravity that caused the initial collapse into the collections of matter as stars and galaxies.

The belief that the observed redshifts of galaxies can be interpreted as the result of receding velocity and the Doppler effect and that the galaxies are thus receding and the universe is expanding is wrong and misleading. In my opinion, this is wrong because there are no direct observations of the receding velocity of galaxies and that there are other causes of redshift. An example of other causes of redshift is the gravitational redshift observed in the spectrum of light emitted from the sun.

In another essay it will be shown how gravitational drag or friction by interstellar gas and dust can reduce the energy of photons as a linear function of distance without involving actual collisions that would otherwise result in blurred images of the photon source.

The following is an analysis of reported observations to show that the universe is closed, and is provided to identify observations that demonstrate that the universe is closed.

There is much concern in determining if the universe is open, flat, or closed. An open universe grows in size, a flat universe does not change, and a closed universe decreases in size. This has been of concern to astrophysicists, cosmologists, and other interested parties. Actually any changes will take billions of years, much greater than the lifetime of people, and probably of civilization, and will not actually matter. By answering or avoiding this question, scientists and others can be free to address more serious current problems described in other essays.

If the cosmos is closed, then the individual matter dispersed in the cosmos billions of years ago will start to collect due to gravity into larger collections such as stars and galaxies that can be detected and viewed by the light emitted and that reaches the observer.

As billions of years pass, the collections of visible stars will increase in number because gravity will sweep in local amounts of invisible and disbursed interstellar dark gas and dust to produce additional visible stars and galaxies. Because light travels at a constant speed, we can look back in time for remote distances to see the increase of visible collections of previously disbursed gas and dust, now in the form of visible stars and galaxies.

Thus we can answer the question of open, flat, or closed universe by observing the number of stars and galaxies visible through light reaching the observer as a function of time as indicated by increasing redshifts and the limit by the velocity of light. By observing back billions of years and seeing the increase in the number of collections of visible matter at different times, it will be possible to show that the matter in the cosmos has been collapsing into collections of visible matter. The compacting of matter in the cosmos is part of the process of the closing of the cosmos

The redshifts observed by Hubble as a linear function of distance can be used to take observations back in time. The notation for redshifts uses z as the representation of the change in wavelength divided by the wavelength without redshift. An increasing z indicates observations further back in time.

However there are problems in using redshift. Hubble suggested that redshift is due to the Doppler effect and that it is due to the receding velocity of the galaxies, but later expressed doubt about this. We can ignore the velocity speculative interpretation because there were no direct observations of receding velocity. There is a more likely cause of the redshift which is due to the effect of gravity in reducing the energy of traveling photons, resulting in a redshift. This can be observed in the redshift of light emitted from our sun.

A complication is for the need to avoid using most quasars which are really galaxies containing black holes, which will contribute to the redshift and make these quasars, look farther away than they really are.

In documents in the list of interesting publications there is a discussion of galaxy formation for different redshifts z. Because redshifts of galaxies were observed to be correlated with the distance from the observer, (and because the information provided by light travels at a constant speed), the redshift permits the observer to look back to different times in the cosmos.

The amount of star formation at a redshift of $z = 3$ appears to be 10 times below that at redshift $z = 1 - 1.5$. And at redshift of $z = 4$ the star formation is a factor of 5 less relative to $z = 3$.

These observations show that normally invisible matter, such as dust and gas, in the universe will gather in a massive collection of matter that, under gravitational compression, will increase in temperature and become hot enough to emit light. This mass can continue to radiate visible light when nuclear processes then start.

There is published a wedge-shaped map of the observed cosmos where the radial distance is measured in z (redshift and time) which shows that the number of observed collections of matter (galaxies) increases for low z, closer to current time.

The process of increasing star formation as the observation distance shown by reduced values of redshift z and approaching current time shows that with time, the matter in the universe is collecting in bodies of matter, such as stars, galaxies, and cosmic structures such as strings and walls of stars. This is a sign of a collapsing closed cosmos

In the book by Hoyle, Burbidge, and Narlikar, *A Different Approach to Cosmology*, there is a chapter ("The Large Scale Distribution of Matter") that shows the increase in visible matter as a function of redshift z with less visible matter apparent at higher redshift or earlier times.

In that chapter, there is mention of dark matter which in essays in this present book has been shown not to be needed because of the additional gravity term for Newton's gravitational constant that is needed to include observations at cosmic distances.

There is also mention of observed periodicities in the structure of about a scale of about 20 Mpc. Because this observation is in conflict with current theory, the popular belief is that the observations are not believable. However observations of redshifts also show periodicities, probably due to gaps in travel distances (and not due to discontinuities in travel velocity). This supports the observations of distribution periodicities.

Recently, one of the books in my collection was found to contain interesting information about the progressive appearance of visible light from stars that increased with time and with decreasing redshift. This shows that the collection of invisible gas and dust in the early cosmos had collected into clumps of matter by gravity and with the passage of time.

An interesting figure, 4.1 on page 52 of the book *Our Cosmic Habitat* by Martin Rees (2001), shows the "Large scale clustering of galaxies: slices in the Northern and Southern Hemispheres as mapped with the Anglo-Australian Telescope." The observations cover the period back to about 2.50 billion light years and up to a redshift of about 0.3.

This helps to confirm, by observations that the cosmos is collecting into lumps of matter and that the cosmos is closed.

I suggest that this increase in clumps of matter with passage of time deserves additional study by others.

Related to this subject of collection of masses is the observation that galaxies collect in strings and form webs of galaxies. You can see an explanation of this observation in a chapter (page 35) in my previous book "The Misunderstood Universe

---

1. Bothun, Gregory. 1998. *Modern Cosmological Observations and Problems.* pp. 217-218.
2. Steidel, C. et al. 1996. *Astrophysical Journal Letters* 462, L17.
3. Hoyle, F., G. Burbidge, and J. V. Narlikar. 2000. *A Different Approach to Cosmology.* New York: Cambridge University Press.
4. Rees, Martin. 2001. *Our Cosmic Habitat.* Princeton, NJ: Princeton University Press.

Essay 17

How to Understand the True Cosmos

Theories and models are based upon observations. Validated observations must take priority over theories. Even if theories agree with current observations, when additional observations do not agree with a theory, this theory must be modified, extended, replaced, or abandoned.

As an example, Newton's universal law of gravity is only based upon observations in our solar system. This is because in Newton's time, the observations for fixed stars at cosmic distances were not available to Newton. When additional observations were obtained for galaxies at cosmic distances, Newton's law of gravity failed to explain these new observations unless a mysterious amount of invisible matter, dark matter, is introduced to provide the necessary additional gravity. Decades of expensive search by cosmologists—including expensive massive machines—have been unable to locate the missing matter. I predict that the failures will continue.

There is a simple explanation for the new observations at cosmic distances based upon observations reported by Vera Rubin for the rotation of stars in spiral galaxies. These observations show that the stars in the arms of the spiral galaxies rotate at a constant velocity as a function of distance from the center, rather than at decreasing velocity with distance as in our solar system.

The equations balancing the gravitational attraction force, $M^*G=1/r^*r$, and the centrifugal force of rotation, v^*v/r, results in an equation $M^*G=v^*v^*r$. Thus, because of the observed constant velocity v, in the spiral galaxy, the product M^*G is a linear function of distance r.

Rather than assigning the distance dependence to the mass M resulting in the need for dark matter, my Theory of Additional Gravity (TAG) assigns the distance dependence to gravity G, which is already invisible and removes the need for the mystery of dark matter.

Thus the gravitational constant G in Newton's law of gravity is now $G=Gn+A^*r$, where A is a constant multiplier for distance r. For small distances r the value for G reduces to Newton's value Gn and works for observations in our solar system—while for large cosmic distances, the new component A^*r dominates and can explain cosmic observations.

This TAG explanation is different from the interesting MOND theory of Martin Milgrom, which involves acceleration.

The situation for the wrong meaning (Doppler effect and velocity) of redshift is more serious because it results in many wrong beliefs and errors, including the dogma of receding galaxies, the expanding universe, the wrong age of the universe, the big bang, the need for inflation, and the mystery of dark energy. These are explained in my other essays.

A much more serious error in the current understanding of the universe is related to the work of Edwin Hubble in the early 1900s on the observed redshift of galaxies as an increasing linear function of distance. He proposed the Doppler effect due to receding velocity as the cause of the redshift, but later expressed doubt about this interpretation. However most cosmologists used the concept of receding velocity to explain redshift, and this led to the damaging speculation that the galaxies were receding, followed by the belief in the big bang. Based on the belief in the receding velocity and observations of very remote supernovae type Ia, the concept of the expanding universe, the apparent accelerated expansion, and the mystery of dark energy were introduced. Also the observed cosmic microwave background (CMB) was interpreted as proof of the big bang. Nobel Prizes were awarded for work based upon the speculation involving the receding galaxies.

Actually, there were no direct observations of the receding velocity of the supposedly receding galaxies.

There is another likely cause of the redshift, and this is due to the effect of gravity on photon energy (redshift). An example is the known redshift of photons emitted from the sun due to the effects of gravity.

My analysis suggests that gravitational drag by gas and dust in interstellar space on photons travelling long distances to the observer will extract energy from photons by gravitational interactions without needing collisions. The gas and dust interacting gravitationally with the photons will be left with motion energy after the photon passes away. Because the velocity of light is constant, the energy extractions are from the photon vibrational energy resulting in redshifts.

Essay 18

Some aspects of the cosmos are now explained

Theories and speculations are based upon past observations and are confirmed by additional observations. When not confirmed, they must be replaced or modified.

This collection of essays will examine the various mysteries in the Standard Model of the universe along with other important beliefs and will provide alternative theories based upon observations. Speculations will be avoided.

Starting in the early 1900s, with the new access to bigger and more powerful telescopes, many observations of the cosmos outside our solar system became available.

One important example was the determination by Edwin Hubble that the fixed stars in the sky were actually galaxies, each with billions of stars and at very great distances.

When Newton's law of universal gravity (based upon observations of planets in our solar system) and the motion of galaxies in groups were observed by the very capable Fritz Zwicky, the observations could only be explained by additional gravity from supposed massive amounts of invisible matter (dark matter). In other essays in this book, the law of gravity will be extended to also be valid at cosmic distances.

Another example was the important observation by Hubble that remote galaxies showed redshifts that apparently increased as a linear function of distance.

Unfortunately, he also suggested that the redshift was due to the Doppler effect, although he later expressed doubt about this suggestion. Belief (speculation) in the Doppler Effect led to the belief that the galaxies were receding, thus leading to the belief in the big bang.

These essays will explain dark matter and dark energy, show the true cause of redshift, show that it is not caused by the receding velocity of galaxies, show that the age of the cosmos is wrong, explain tired light, solve Olbers's paradox about the dark sky, remove the need for inflation, explain the cosmic microwave background (CMB), understand quasars and black holes, and explain why the cosmos (universe) is apparently closed.

Essay 19

Observations take priority over speculations

The existing model of the universe is full of theories, mysteries, and harmful speculations. However, there are a large number of observations. If a theory cannot be confirmed by prior or subsequent observations, then the theory should be abandoned, replaced, or corrected. Remember, observations must take priority over theories and speculations.

In the current dogma, there are many speculations not confirmed by valid observations.

This collection of essays about cosmic gravity follows the earlier self-published book, *The Misunderstood Universe*" © 2009 which resulted after several decades of studying part-time the many mysteries, errors, and speculations in the Standard Model of the universe. A major mystery is related to dark matter and another is related to dark energy, both of which can be explained by my concept of cosmic gravity. Serious errors related to redshift and the beliefs in its associated velocity are also disclosed.

After several years of additional consideration, I realized that there are two major causes for the errors in the understanding of the current dogma about the cosmos (universe), both related to the use of Newton's universal law of gravity and Newton's gravitational constant G for observations at cosmic separations, and thus outside of the solar system that provided the observations used by Newton.

One cause of the errors is the assumption by many others that Newton's universal law of gravity and his gravitational constant Gn are also valid at cosmic distances, although only based upon observations of planets in our solar system.

The second and more serious cause of error is the additional observations by Edwin Hubble who, in the early 1900s, used a more powerful telescope to study the fixed stars in the sky and determined that they were galaxies located at vast distances and consisted of hundreds of thousands of millions of stars.

He also measured the redshifts of the remote galaxies and determined that the redshifts were a linear function of their distance from the observer. Unfortunately he speculated that the redshifts were due to the Doppler effect and that the galaxies were receding. He later expressed doubt about the Doppler effect interpretation, but others accepted this interpretation as fact.

Note that there are no direct observations of the receding velocities.

The serious consequences of the speculative interpretation of redshift include the following errors and questionable beliefs that would require abandoning some well-known laws of physics:

(1) The wrong age of the cosmos (based upon the Hubble constant when expressed in velocity terms)

(2) Belief in the receding galaxies

(3) Belief in the expanding cosmos

(4) Belief in the big bang

(5) The need of inflation for the observed uniformity of the Cosmos beyond the event horizons

(6) Belief that in the first fraction of a second of the supposed big bang the *laws of physics* relating to expansion velocities greater than Einstein's velocity of light could not apply

(7) Belief that in the first fraction of a second of the supposed big bang the *laws of physics* relating to gravity could not apply and that the expansion could escape the gravitational attraction of the initial compact massive energy/mass;

(8) Belief that the cosmic microwave background (CMB) and its very low temperature was due to the cooling of the initial high temperature at the start of the speculated big bang;

(9) Belief that the supposed velocities of very remote galaxies had accelerated compared to their location as determined by their

observed redshifts (as a function of distance), thus requiring dark energy to power the acceleration;

(10) Suggestions that the supposed massive dark energy could supply much of the missing dark matter according to Einstein's $E=m*c*c$.

The introduction of cosmic gravity as a solution to the mystery of dark matter can result from the realization that Newton died before Edwin Hubble (in the early 1900s) had used the more powerful telescopes then available to obtain observations for the very remote galaxies and stars in the cosmos. Thus Newton only could use observations for planets in our solar system for his law of gravity and his gravitational constant, Gn.

Fritz Zwicky, in my opinion, was an excellent scientist and important contributor to the concept of supernovae and an important contributor to the collection of observations. Later in his career, he was treated very badly by the astronomy establishment and deprived of observation access. Zwicky used Newton's gravitational constant to study the observed motion of groups of galaxies in the cosmos and he found that the observations could only be explained by large amounts of missing matter. This was an early indication of the need for dark matter, and was based upon the common existing belief (speculation) that Newton's constant also applied outside the solar system

Decades later, Vera Rubin studied the velocity rotation curves in the arms of spiral galaxies and determined that the velocities were flat (constant) as a function of radial distance rather than decreasing as in our solar system. She then suggested that this was due to a halo of massive dark matter for the spiral galaxy.

After I reviewed data for Newton's law of gravity and included the cosmic observations along with the planetary observations for our solar system, I was able to extend Newton's gravitational constant Gn to include the cosmic observations for cosmic distances r. The result is the gravitational constant in the new cosmic gravity form of Gs as now $Gc=Gn+A*r$ where A is a constant determined from observation of rotation of stars in spiral galaxies by Vera Rubin.

At first when I learned of the observations of Rubin concerning the flat, constant rotational velocities in spiral galaxies, it became apparent to me that the constant velocities could be explained if the gravitational constant of Newton and the inverse square of radius r dependence of the force could be replaced by an inverse r dependence. I then suggested that for larger values of r the gravitational constant included a term linear in r which was the first term of a series expansion for G. This would explain and predict the additional force that would become dominant when the inverse square distance force became much smaller.

Several years later, I realized that the linear term without approximations or series expansion comes directly from a basic physics equilibrium equation that balances the basic gravitational force against the centrifugal force of rotation.

When the force of gravity F=M*G/r*r is balanced against the force of centrifugal motion F=v*v/r, the result is M*G=v*v*r. (Note that the symbol * used here represents the multiplication operator.) For the case where the velocity v is constant, the product M*G is a linear function of distance r. Instead of assigning the distance dependence to mass M as is usually done leading to the fruitless search for dark matter, I assign the distance dependence to gravity G and remove the need for dark matter. Note that gravity and the supposed dark matter both have similar properties. They both are not visible, do not reflect light, do not eclipse light, and are only apparent by their effect on visible matter.

Observations by Rubin and others for spiral galaxies indicate that the term A*r becomes dominant for distances r greater than about 2.7 kpc (kilo parsecs). For smaller values of distance r (such as in our solar system), the cosmic gravity constant Gc *without change* reduces to the usual Newton's gravitational constant Gn and continues to explain solar observations. Note that this term A*r for small distances in our solar system can explain the NASA Pioneer anomaly for the space probes 10 and 11, when A*r becomes smaller by many orders of magnitude for the small solar system values of r.

Theories and speculations are based upon past observations and must be confirmed by additional observations. When not confirmed, they must be replaced or modified.

While I am not trained as a cosmologist, I have been able to quickly and productively work in many fields and disciplines. My credentials as a scientist are presented in Appendixes

Essay 20

Some comments

A great deal of energy is being devoted to finding Earthlike planets in galaxies at cosmic distances. We should realize and admit that there is no current possibility of traveling to or from such planet within the relatively limited lifetime of current humans, even if travel at the speed of light is possible or practical.

We cannot even communicate with any intelligent life on any livable planet light years away, unless the interchange is acceptable for delays over multiple generations.

The very wild speculation about travel through wormholes will probably not be shown to be possible during the expected limited duration of our civilization. It may not be possible at all.

We are now going through a period of global warming coming out of a major ice age about 10 million years ago and a little ice age about 600 years ago. These recoveries from ice ages are more powerful and probably more important in influencing the temperature of the Earth than any effects of greenhouse gasses.

At some point in the far future we can expect another ice age and hope that our technology advances enough so that we can mostly survive.

The way our population is outgrowing our resources, at some point, our population growth in future generations must slow or reverse.

Our stored energy resources for the world (oil, gas including shale gas) will be limited to several centuries but eventually will run out. Even nuclear fuel will be limited. We eventually must survive using contributions from renewable energy from our sun (solar, wind, water) as long as the sun exists.

A very promising energy source that must be developed in spite of resistance by entities—companies and governments—that would be harmed economically or politically when a practical means of fusion of hydrogen or other elements is demonstrated and developed. The solution probably will involve a form of cold fusion.

If we can slow or divert the waste of some funds and of some scientists and engineers on wrong work on cosmology resulting in wrong beliefs, the funds and work can be applied to more critical problems such as improved health, energy resources, food resources, water resources, weather problems, climate changes, earthquakes, and other problems. Many of our scientists and engineers are versatile and could be productive in meeting the new objectives.

Speculation about travel through worm holes or multiple universes or time travel or other theories will eventually require scientific demonstration.

The suggestion that redshift and the increase in photon wavelength are due to the expansion of space-time is really not reasonable because space-time is just a mathematical description used to model some observations and apparently cannot be proven to exist in a physical form.

The concept of space-time, involving the dimensions X, Y, and Z as well as time, is just an interesting mathematical theory and model not supported by direct observations.

Essay 21

Why science and civilization failed, and suggestions for critical problems

I believe that science and civilization has failed to benefit and protect future generations.

Mankind and scientists in particular are strongly motivated to understand their environment in the past, present, and future. They are driven by curiosity, survival, career, and need for recognition.

Civilization and science have failed in truly understanding the cosmos, and starting in the early 1900s, with the availability of more powerful telescopes giving access to observations of stars and galaxies at cosmic distances, much time and money was wasted in speculations about the meaning of these new observations.

The funding by government agencies and private sources can perhaps be replaced, but the passage of time cannot be stopped, and we are coming closer to serious problems that can threaten civilization.

In the earlier centuries, the cosmos was of interest to many people who observed the fixed stars and the moving stars (planets). They used their time and, in some cases, their funds to make and use early telescopes to study the sky. Sometimes their work was supported by royalty and by wealthy families.

Experts in the field of cosmology continued with existing beliefs and dogmas and apparently resisted the contributions of others such as Zwicky and Arp who were denied access and time on modern telescopes.

Because of the increased costs associated with more powerful telescopes, federal funding is a requirement for construction and operation of these telescopes, including those expensive ones operating in space outside our atmosphere.

Funds for support of scientists and engineers and their projects are also obtained from foundations, company funds, and some wealthy individuals.

Some of the early results describing the solar system were obtained by wealthy people who built their own telescopes in order to observe the planets and moons in our solar system. It was only after much larger telescopes were available, such as the large one used by Hubble, and funded by organizations, it was possible for Hubble to view the fixed stars and determine that they consisted of galaxies at cosmic distances.

Awards and grants for results are important for continued support of projects and associated scientists and engineers. Universities, contract companies and investigators depend upon continued support

There are many problems and mysteries left to consider for the benefit of future generations.

Scientists and engineers can still solve some of these problems, and some problems need political power and funds that can be supplied by federal funding, foundations, successful business people, and successful scientists and technologists.

The following are some suggestions for problems to be considered:

ENERGY

* Finding new practical sources of energy.
* Finding new replacements for petroleum.
* Devising better storage batteries.
* Solving the problem of convenient room temperature superconductors or higher conductivity wires for the electric power grid.

HEALTH

* Providing means to control or limit viral/bacteria epidemics.
* Reducing the cost of health care.
* Creating new and better means to improve the health of people.
* Providing health care and insurance for everyone and that is more effective and cost-controlled.
* Developing more effective ways of controlling or curing cancer.
* Developing better and less expensive ways of controlling infections, such as bacteria and viruses.
* Devising improved stents to improve blood circulation.
* Devising practical ways of assisting hearts while solving the limitations of tissue and blood compatibility.

WATER

* Solving the need for new sources of water.
* Providing water for the parched regions, possibly by desalinization of ocean water, using unlimited energy.

UNIVERSE

* Arranging to detect and prevent asteroids from impacting the Earth with enough force to destroy most life, as has happened in the past.

SURVIVAL

* Recognizing and determining if significant global warming really exists.
* Providing means to survive global warming if it is true.

* Realizing and showing that the sun itself has actually determined the temperature of the Earth in the past thousands of years, before the effects of man.

* Realizing that the Earth is warming because we are coming out of the last ice age and the little ice age.

* Showing that CO2 may be accelerating the warming of the Earth, but realizing that reducing this increase in CO2 will only delay the warming for only the present generation.

* Sequestering the excess CO2 in plants as a more efficient way of recycling it together with the sunlight stored in plants for a future energy resource, as was done by nature in past eons.

* Developing essentially free energy to power the means for us to survive the higher and lower temperature periods including the next ice ages.

* Developing crops that are more productive and efficient and free of patent control by agricultural seed companies.

GENERAL

* Ensuring that members of Congress represent the true wishes of the country rather than being controlled by lobbyists or giving priority for their reelection over the needs of the citizens and voters.

* Devising ways of controlling hackers and crimes on the Internet.

* Improving the means for transportation.

* Reducing crime.

* Improving the Internet defense capabilities of companies and of countries.

* Providing survival of civilization in the event of significant global warming or cooling.

* Giving people means to protect themselves against guns but without the ability to hurt attackers.

* Protecting the Earth against the destruction of the electrical power network (and of civilization) that would be caused by a giant flare from the sun hitting the Earth. Particularly serious during a long period when the Earth's protective magnetic field

reverses and decreases during the reversal as it has done many times in the far past.

There are many other national problems that the reader can identify.

Essay 22

Summary of these essays

Introduction

There are many mysteries and speculations in the current dogma about the cosmos. One speculation is that Newton's universal law of gravity also is valid in the vast cosmos outside our solar system, leading to the mystery of dark matter. Another much more serious speculation, resulting in errors that are more serious, is due to the belief that Edwin Hubble's redshift is due to the receding velocity of cosmic galaxies, leading to the mystery of dark energy.

We will discuss these speculations and other errors in the current model of the cosmos (universe) and will provide corrections based only upon observations.

This collection of essays involving aspects of cosmic gravity follows my earlier book related to the subject. This book *The Misunderstood Universe* was published in 2009 and considered the many mysteries and errors and speculations in the Standard Model of the universe. In these essays, there is a difference between valid observations and just speculations and theories. We will identify information based upon valid observations. Information based only on speculations will be individually identified as speculations.

After additional years to consider the matter, I decided that a major cause of the mysteries and other problems was related to the fact that the genius Isaac Newton only had available observations of planets in our solar system and did not have access to observations of fixed stars and galaxies at great distances in the cosmos and outside our solar system. This lack of additional observations for galaxies and stars in the cosmos resulted in the belief in dark matter and the fruitless and expensive continuing search for dark matter.

By introducing the concept of cosmic gravity, based upon cosmic observations, I am able to provide a linear extension for Newton's gravitational constant Gn to result in a cosmic gravitational constant Gc that continues to work for observations in our solar system and also works at cosmic distances. This will remove the need for dark matter.

A more serious error was the result of the work of Edwin Hubble in observing that there are redshifts for galaxies in the cosmos, but speculated that the redshifts are due to receding galaxies. This resulted in many more serious speculations that led to distorted views of the cosmos.

This collection of essays is intended to correct serious errors in the current model of the universe and to provide a correct understanding of the cosmos. It is primarily intended for future generations of scientists and for many others interested in understanding the cosmos. Probably many in the current generation of scientists (because of career objectives, ego, and need for continued financial support), will resist this corrected understanding and stick to the current dogma, as has hampered scientific progress in past generations.

However, after several decades of studying part-time the mysteries in the standard model of the universe, I realized that many of the mysteries and speculations in the current dogma can be explained and removed.

They are caused by assuming that Newton's universal law of gravity also applies to observations at cosmic differences, far outside our solar system. Newton died before the work of Edwin Hubble (in the early 1900s) who used more powerful telescopes to observe the fixed stars.

Hubble determined that the fixed stars are galaxies located at vast distances and that each consists of hundreds of thousands of millions of stars. The genius Newton did not have knowledge of the observations for the very remote stars in the cosmos, so he only used the available observations for planets in our solar system, and his theory of gravity resulted in many mysteries and speculations in the present dogma about the cosmos.

When I added the new cosmic observations with the planetary observations for our solar system, then Newton's law of gravity and Newton's gravitational constant Gn was extended to include the cosmic observations for cosmic distances r to be in the new cosmic gravity form of the gravitational Gc as Gc=Gn+A*r where A is a constant determined from observation of rotation of stars in spiral galaxies by Vera Rubin.

Observations by Rubin and others indicate that for spiral galaxies in the cosmos, the rotation velocity curves become flat for distances r greater than about 2.7 kpc (kilo parsec)

It will be shown in later essays that the equations balancing the gravitational force and the centrifugal rotational force reduces to M*G=r*v*v. For the case where the observed velocity v is constant, this requires that the product of the mass M and the gravitational constant G is a linear function of distance r. Rather than, as usual, assuming that G is constant and that the mass M has the distance dependence (thus requiring dark matter to supply the associated gravity), we instead assign the linear distance dependence to the already invisible gravity term G, thus removing the need for dark matter.

Note that the supposed dark matter has the properties of not emitting light, not reflecting light, nor eclipsing light from other sources, but is capable of influencing the motion of observable entities. Gravity, which is admitted to exist already, has the same properties and can explain the new observations in the cosmos without the need to search for dark matter.

Thus the gravity term in the cosmos Gc can now be in the form of Gc=Gn+A*r where Gn is Newton's gravitational constant and A is a constant that describes where the term A*r becomes dominant for distances r greater than about 2.7 kpc (kilo parsec), which is the distance where the rotation velocity curves as a function of radius become flat.

For smaller values of distance r (such as in our solar system) the cosmic gravity constant Gc reduces to the usual Newton's gravitational constant Gn.

Observations take priority over theories and speculations, and theories must be confirmed by valid observations. When not confirmed, they must be replaced or modified. In my opinion, the present dogma about the cosmos is full of theoretical speculations that reduce the value of the many careful and valuable observations.

In our analysis in this second book containing the new collection of essays about the cosmos and involving cosmic gravity, we will consider in more detail the speculations and mysteries including some of the following: cosmic gravity; dark matter; Hubble and fixed stars; redshifts; the age of the universe; receding galaxies; the expanding universe; the big bang; Inflation; CMB (cosmic microwave background); quasars; black holes; dark energy; standard candles (supernovae type Ia or SN Ia); Olbers' paradox; the Pioneer anomaly; and an answer for the open, flat, or closed universe. Two of the more recent Nobel Prizes in Physics (1978 and 2011) should be reconsidered in view of the additional information presented here.

Some of the essays of concern will be outlined below.

Cosmic gravity

Newton's universal law of gravity is only based upon observations in our solar system because in Newton's time, the observations for fixed stars at cosmic distances were not available to Newton. When additional observations were obtained for galaxies at cosmic distances, Newton's law of gravity failed to explain these new observations unless a mysterious amount of invisible matter, dark matter, is introduced to provide the necessary additional gravity. Decades of expensive search by cosmologists using expensive massive machines have been unable to locate the missing matter. I predict that the failures will continue. The expectation that the recently introduced short-lived Higgs boson will supply the missing mass will probably be found to be disappointing.

The need for dark matter to explain observations in the cosmos (involving the fixed stars that are actually galaxies) is due to the inability of Newton's law of universal gravity to explain the cosmic observations.

There is a simple explanation for the new observations at cosmic distances based upon observations reported by Vera Rubin for the rotation of stars in spiral galaxies. These observations show that the stars in the arms in the spiral galaxies rotate at a constant velocity as a function of distance from the center, rather than at decreasing velocity with distance as in our solar system.

The equations balancing the gravitational attraction force,

$$F = M{*}G/r{*}r \qquad (1)$$

and the centrifugal force of rotation,

$$F = v{*}v/r \qquad (2)$$

results in a balancing equation

$$M{*}G = v{*}v{*}r. \qquad (3)$$

Thus because of the observed constant velocity v in the spiral galaxy, the product M*G is a linear function of distance r.

Rather than assigning the distance dependence to the mass *M* resulting in the need for dark matter, my introduction of the new concept of cosmic gravity or the earlier name of the Theory of Additional Gravity (TAG) assigns the distance dependence to gravity G which is already invisible, and removes the need for the mystery of dark matter.

Note that gravity has the same properties as the supposed dark matter: they both are not visible or do not emit light, do not reflect light, nor do not block light, and both are only apparent by influencing the motion of visible objects.

Thus according to my new analysis, the gravitational constant *Gn* in Newton's law of universal gravity is now the cosmic gravity constant Gc defined as:

$$Gc = Gn + A{*}r \qquad (4)$$

where A is a constant multiplier for distance r.

For small distances *r* the value for Gc reduces to Newton's value Gn and works for observations in our solar system, while for large cosmic distances, the new component A*r dominates and can now explain cosmic observations.

This cosmic gravity or TAG explanation is different from the well-known MOND theory of Martin Milgrom, which involves acceleration and apparently has limitations.

Dark matter

Newton's universal law of gravity is only based upon observations in our solar system because in Newton's time, the observations for fixed stars at cosmic distances were not available to Newton. When additional observations were obtained for galaxies at cosmic distances, Newton's law of gravity failed to explain these new observations unless a mysterious amount of invisible matter, dark matter, is introduced to provide the necessary additional gravity. Decades of expensive search by cosmologists (including expensive massive machines) have been unable to locate the missing matter. I predict that the failures will continue. The suggestion that the recently detected very short-life Higgs boson could supply the missing mass will probably not be valid.

The need for dark matter to explain observations in the cosmos (involving the fixed stars that are actually galaxies) is due to the inability of Newton's law of universal gravity to explain the cosmic observations.

When Newton's law of universal gravity (only based upon observations of planets in our solar system) and when the motion of galaxies in groups were observed by Fritz Zwicky, the observations could only be explained by the additional gravity of from supposed massive amounts of missing invisible matter (dark matter). The law of cosmic gravity will be extended to also include observations at cosmic distances in order to also be valid at cosmic distances.

Hubble and fixed stars

Starting in the early 1900s, with new access to bigger and more powerful telescopes, many observations of the cosmos outside our solar system became available.

One important example was the determination by Edwin Hubble that the fixed stars in the sky were actually galaxies each with hundreds of thousands of millions of stars and are apparently fixed stars because of their very great distances.

Redshifts

Much more serious errors in the current understanding of the universe are related to the work of Edwin Hubble in the early 1900s on the observed redshift of galaxies as an increasing linear function of distance. He proposed the Doppler effect due to receding velocity as the cause of the redshift, but later expressed doubt about this interpretation. However most of the cosmologists used the concept of receding velocity to explain the redshift, and this led to the damaging speculation that the galaxies were receding, followed by the belief in the big bang. Based on the belief in the receding velocity and observations of very remote standard candles (supernovae type Ia), the concept of the expanding universe, the apparent accelerated expansion, and the mystery of dark energy were introduced. Also, the observed cosmic microwave background (CMB) was interpreted as proof of the big bang. Nobel Prizes in Physics were awarded for work based upon the speculation involving the receding galaxies, the expanding universe, and the big bang.

A result of Hubble's work was the important observation by Hubble that the remote galaxies showed redshifts that apparently increased as a linear function of distance. Unfortunately, he also suggested that the redshift was due to the Doppler effect, although he later expressed doubt about this suggestion.

Belief in the Doppler effect led to the belief that the galaxies were receding, thus leading to the belief in the big bang. Our analysis will show the true cause of redshift and will show that at cosmic distances, they are not caused by the receding velocity of galaxies.

The situation for the wrong meaning (Doppler effect and velocity) of redshift is serious because it results in many more wrong beliefs and errors, including the dogma of receding galaxies, the expanding universe, the wrong age of the universe, the big bang, the need for inflation, and the mystery of dark energy. These are explained in my other essays.

Actually, there were no direct observations of the receding velocity of the supposedly receding galaxies.

There is another likely cause of redshift and this is due to the effect of gravity on photon energy (redshift). An example is the known redshift of photons emitted from the sun due to the effects of gravity.

My analysis suggests that gravitational drag by gas and dust in interstellar space on photons travelling long distance to the observer will extract energy from photons by gravitational interactions without needing collisions. Because the velocity of light is constant, the energy extractions from the photons are from the photons' vibrational energy, resulting in increased wavelengths.

The longer range of the new cosmic gravity means that gas and dust at farther distance will also contribute to the loss of photon energy and redshift.

The lost photon energy is now in the motion of interstellar gas and dust which were moved by gravitational interactions with the travelling photons.

This is similar to the transfer of energy from our moon to our Earth in the form of tidal motion without requiring physical contact. As a result, the rotation velocity of the moon decreases, and the orbit of the moon distance increases.

Age of the universe

When the Hubble constant is specified in terms of velocity rather than distance, the reciprocal of the Hubble constant has the dimensions of time and is used to specify the age of the universe.

Because the Hubble constant is *not* related to the Doppler effect or velocity, then the Hubble constant in the velocity form cannot be used to determine the age of the universe. Also, it is observed that some stars are apparently older than the age derived from the Hubble constant.

Receding galaxies

The speculation that the Hubble constant is related to velocity has resulted in the speculation that galaxies are receding and other consequent speculations.

Expanding universe

The speculation that the galaxies are receding has led to the speculation that the universe is expanding and other speculations, including the speculative big bang.

Example of circular reasoning

Redshift is used to demonstrate that the universe is expanding. Surprisingly, others have explained that redshift is due to the expansion of the universe. This is not logical—because A depends upon B, and B depends upon A.

This is an example of the errors and speculations present in the current dogma about the universe (cosmos) and a reason for the need for professional corrections.

Big bang

The speculation about the expanding universe has led to the speculation about the big bang. The desire of some to use the concept of the big bang to support religious beliefs is similar to the effect of religious dogma on scientific work as in past centuries.

Inflation

The observed uniformity of the cosmos as far as can be observed, in conjunction with the belief in the big bang, has to be explained because the event horizon is computed the be much smaller than the size of the universe. The age of the universe and the limit of the velocity of light determine the event horizon, and there is not enough time for interaction of the regions of the cosmos in far-separated event horizons to become equal or similar as the cosmos expands from a small singularity in a speculated tiny fraction of a second.

The answer proposed by others is that when the cosmos was very tiny, it equilibrated and then it inflated to its cosmic size while retaining its uniformity.

Inflation is not needed if we recognize that the Hubble constant does not represent receding velocity, or the expansion of the cosmos, nor does it support speculation about the big bang.

Even the concept of inflation is not needed to explain the uniformity of the cosmos. Consider the case of three event horizons, A, B, and C, where B and A slightly overlap and B and C also slightly overlap. Then B = A and B = C. The logical result is also that A = C. Thus A, B, and C are all equal and that the parts of the cosmos that they contain are uniform.

This logic can be extended to be used in all regions of the cosmos and the speculative concept of inflation is not needed.

CMB (Cosmic Microwave Background)

It was observed that microwave energy (at about 3 cm, X-band) was received from all directions in the Cosmos and it was speculated that it was the cooled remains of the big bang which was speculated to have occurred billions of years ago and started at a very high speculated temperature. Subsequent observations found that the CMB was that of a source in thermal equilibrium and at a temperature of about 2.7 K. Note that a Nobel Prize was awarded in 1978 to the CMB discoverers, Penzias and Wilson. (See *www.nobelprize.org*.)

Actually, we have indicated that the redshift for cosmic distances does not show the speculated receding velocity of galaxies. Thus there are no observations showing the speculative expansion of the cosmos and the subsequent speculation about the big bang.

Quasars

Quasars are galaxies that emit large quantities of RF energy and because of their observed large redshifts are speculated to be very far away. Based upon their brightness and the large distance determined from their redshifts, the energy emitted by quasars is computed to be extremely large. Also, their proper motion (angular transverse velocity in degrees per unit time) based upon their apparent distance is computed as larger than the Einstein limit of c.

Both of these surprising results can be explained by the presence of a black hole or large collection of mass in the quasar, which, because of the gravitation redshift of emitted photons, will result in the apparent distance of the quasar to be much larger than actual. When used to calculate the energy output and the linear proper motion, the incorrect distance determined by the observed redshift will result in incorrect values for the deduced energy output and the linear proper motion.

The large mass in the quasar will also explain the RF energy emitted. Photons climbing out of the large gravitational well in the quasar will have their energy reduced to that of the observed RF energy and may contribute to the observed RF energy.

Black holes

Black holes are massive collections of matter where the gravitational attraction is so great that stars, galaxies, gas, and dust is attracted, and when they enter the gravitational well, it is difficult to escape. There is a resulting problem in connection with the law of conservation of information because presumably, the information drawn into the black hole cannot escape. The genius Stephen Hawking, along with Penrose and others, has studied the problem of the information paradox, and recently, Hawking has admitted that some of his solutions may be wrong and at a meeting has suggested the possibility of a speculative solution related to multiple universes.

An alternate solution for the information paradox is suggested here. It is that the material in the black hole is in thermodynamic equilibrium with a Maxwell-Boltzmann energy distribution. This means that some of the photons are in the high-energy tail of the energy distribution and can escape over the gravitational energy barrier in the black hole. They leave as low energy photons (redshifted radio energy). The emitted photons carry information (similar to radio and TV signals), and this can solve the information paradox.

Dark energy

When Perlmutter, Schmidt, and Riess and their teams observed that some very remote galaxies were much farther away in the cosmos than expected from their redshifts, it was speculated that they had accelerated, and the speculated additional energy required for the speculated receding motion and the speculated acceleration was called dark energy. This is surprising because the distance for very remote galaxies (and their standard candles SN Ia) as determined from their optical magnitude was used to extend the redshift vs. distance calibration relationship.

In any event, the speculated dark energy plus Einstein's relationship between energy and mass was speculated to supply some of the missing dark matter.

A Nobel Prize in Physics was awarded in 2011 to Saul Perlmutter, Brian P. Schmidt, and Adam G. Riess for their work related to dark energy. Because the receding motions of the remote galaxies are only based upon the speculation that redshift is due to the Doppler effect, the basis for the accelerated expansion and dark matter is apparently wrong. However, the very comprehensive measurements by Perlmutter, Schmidt, and Reiss may be worth the Nobel Prize. (See *www.nobelprize.org*.)

Standard candles Supernovae type Ia (SN Ia)

In determining distances of stars and their galaxies by their observed magnitudes (light received and the inverse square dependence on distance) of supernovae type Ia are important because their magnitudes are constant within about an order of magnitude. There are two basic types: one with a lower magnitude with faster rise and decay times, and other much larger types with slower rise and fall times. The lower magnitude SN Ia is probably more accurate in use as a standard candle because it probably ignited reaching the Chandrasekhar limit by slower accumulation of matter, while the larger magnitude SN Ia passed the Chandrasekhar limit by collision with a much more massive amount of matter, which could be more variable.

Olbers' paradox

Years ago, Olbers asked a question about why the sky is black at night in spite of the observation that there are many stars in the cosmos. The explanation we offer is that because of the redshift dependence upon travel distance, some of the photons have not travelled far enough to drop into the visible range of humans, while other photons may have traveled far enough to be below the visible range of humans. The longer range photons can be detected with infrared instruments, or with microwave and RF detector arrays.

Pioneer anomaly

Precision longtime observations of the motion of the NASA space probes Pioneer 10 and 11 have shown that there is apparently a very tiny but measurable attraction of the probes toward the central sun. The observed additional central force is about 8 orders of magnitude less than that for Newton's gravitational constant. Alternate explanations such as gas leakage has been considered but apparently ruled out.

Considering the suggested cosmic gravity Gc and the cosmic distance dependence described previously in the form of $Gc=Gn+A^*r$, the Pioneer anomaly can be explained by the very small value of r in the solar system.

Open, flat, or closed universe

The universe can be suggested to be closed (collapsing) due to gravity (including the cosmic gravity contribution) if observations show that the number of observable matter at larger distances, larger redshifts, and earlier times in the cosmos is less than can be observed at closer distances, smaller redshifts, and at more recent times.

Because it takes time for less visible interstellar gas and dust to collect into visible stars, there should be a progressive increase toward the present time for matter in structures of stars to ignite and to appear and become visible.

Appendices

Appendix A

References and additional reading material

This represents some of the material located and read in the process of this analysis. They are now in my personal library. The books were purchased through the Internet and from sources such as Amazon that sell copies of many low-cost books that are out of print. (For help, Google *+amazon +books*.) Information available on the Internet also was very useful.

Anderson, John D., Philip A. Laing, Eunice L. Lau, Anthony S. Liu, Michael Martin Nieto, and Slava G. Turyshev. 1998. "Indication, from Pioneer 10/11, Galileo, and Ulysses Data, of an Apparent Anomalous, Weak, Long-Range Acceleration." *Physical Review Letters* 81, 14: 2858-61.

Anderson, John D., Slava G. Turyshev, and Michael Martin Nieto. 2002. "A Mission to Test the Pioneer Anomaly." *International Journal of Modern Physics*, D. vol. 11, 10: 1545-51.

Arp, Halton. 1998. *Seeing Red: Redshifts, Cosmology, and Academic Science*. Montreal: Apeiron.
(Also *http://redshift.vif.com*. An example of the resistance of established cosmologists to new correct ideas.)

Arp, Halton. 1987. *Quasars, Redshifts, and Controversies.* N.p.: Interstellar Media.

Assis, A. K. T. and M. C. D. Neves. 1995. "History of 2.7 K Temperature Prior to Penzias and Wilson." *Apeiron* 2: 79-84.
(Describes the important independent determination of the temperature of interstellar space.)

Bahcall, N. A. 1997. "Large Scale Structure in the Universe." In *Unsolved Problems in Astrophysics*, edited by John N. Bahcall and Jeremiah P. Ostriker, 61-91. New Jersey: Princeton University Press.

Bartusiak, Marcia. 2004. *Archives of the Universe.* New York: Vintage Books.
(An excellent collection of information that is well worth reading.)

Bothun, Gregory. 1998. *Modern Cosmological Observations and Problems.* London: Taylor & Francis.
(Other modifications of Newton's law have been proposed along with discussions of the many problems in the current cosmological models.)

Cohen, Nathan. 1998. *Gravity's Lens.* New York: John Wiley and Sons Inc.

Ferguson, Kitty. 1990. *Measuring the Universe.* New York: Walker and Company.

Goldsmith, Donald. 1991. *The Astronomers.* New York: St. Martin's Press. pp. 36-44.

Gregory, Jane. 2005. *Fred Hoyle's Universe.* Oxford: Oxford University Press.

Guth, Alan. 1997. *The Inflationary Universe.* Cambridge, MA: Perseus Books.

Harrison, Edward R. 1981. *Cosmology: the Science of the Universe.* Cambridge: Cambridge University Press.

(Page 240 discusses the tired light concept of Zwicky.)

Hawking, Stephen W. 1988. *A Brief History of Time*. New York: Bantam Books.

Hawking, Stephen W. 2002. *The Theory of Everything*. Beverly Hills, CA: New Millennium Press.

Hirschfield, Alan W. 2001. *Parallax*. New York: W. H. Freeman.

Hoyle, Fred, Geoffrey Burbidge, and Jayant Vishnu Narlikar. 2000. *A Different Approach to Cosmology*. New York: Cambridge University Press.

Hubble, Edwin. 1929 "A Relation between Distance and Radial Velocity among Extra Galactic Nebulae." *Proceedings of the National Academy of Science of the United States of America* 15: 168-73.

Hubble, Edwin. 1937, *The Observational Approach to Cosmology* Oxford, England: Clarendon Press. p. 68.

Hubble, Edwin and Milton Humason. 1931. *Astrophysical Journal*. 74, 43.

Hubble, Edwin. (1936) 1982. *The Realm of the Nebulae*. New Haven: Yale University.

Kragh, Helge. 1996. *Cosmology and Controversy*. Princeton, NJ: Princeton University Press.

Kushner, R. P. 2002. *Extravagant Universe*. Princeton, NJ: Princeton University Press.

Lerner, Eric J. 1992. *The Big Bang Never Happened*. New York: Vintage Books.

Levy, David H., ed. 2000. *Scientific American Book of the Cosmos*. New York: St. Martin's Press.

Lightman, Alan. 2005. *The Discoveries*. New York: Pantheon Books.

Livio, Mario. 2000. *The Accelerating Universe* N.p.: John Wiley.

Milgrom, M. October 1998. arXiv: astro-ph/9810302 v1: 20.

Milgrom, M. 1983. *Astrophysical Journal*. 270: 365-70.

Milgrom, Mordehai. 2002. "Does Dark Matter Really Exist?" *Scientific American*. pp. 42-52.

Murdin, Paul. 2009. *Secrets of the Universe*. N.p.: University of Chicago Press.

Narlikar, Jayant Vishnu. 1993. *Introduction to Cosmology* 2nd ed. N.p.: Cambridge University Press.

Peebles, Philip James Edwin. 1993. *Principles of Physical Cosmology*. Princeton, NJ: Princeton University Press.

Perlmutter, Saul. April 2003. "Supernovae, Dark Energy, and the Accelerating Universe." *Physics Today*. p. 53-60.

Perlmutter, Saul. *http://supernova.lbl.gov/PhysicsTodayArticle*.pdf.

Rees, Martin. 2001. *Our Cosmic Habitat*. Princeton, NJ: Princeton University Press.

Rubin, V., D. Burstein, W. K. Ford Jr., and N. Thonnard. 1985 "Rotation Velocities of 16 Sa Galaxies and a Comparison of Sa, Sb, and Sc Rotation Properties." *Astrophysical Journal* 289: 81.

Rubin, Vera. and W. Kent Ford Jr. 1970. "Rotation of the Andromeda Nebula from a Spectroscopic Survey of Emission Regions." *Astrophysical Journal* 159:379.

Silk, Joseph. 2001. *The Big Bang* 3rd ed. New York: W. H. Freedman and Company.

Smolin, Lee. 1997. *The Life of the Cosmos.* New York: Fordham University.

Sofue, Yoshiaki and Vera Rubin. 2001. "Rotation Curves of Spiral Galaxies." *Annual Review Of Astronomy and Astrophysics* 39:137-74. (Also   http://www.physics.ucla.edu/~cwp/articles/rubindm/rubindm. html.)

Spergel, David. 1997. "Dark Matter." In *Unsolved Problems in Astrophysics*, edited by John N. Bahcall and Jeremiah P. Ostriker, 221-40. Princeton, NJ: Princeton University Press.

Tifft, William G. and W. John Cocke. January 1987. "Quantized Galaxy Red Shifts" *University of Arizona Sky & Telescope Magazine.* pp.19-21

Weinberg, Steven. 1988. *The First Three Minutes* 2nd ed. New York: Basic Books.

Zwicky, Fritz. 1929. "Redshift of Spectral Line." *Proceedings of the National Academy of Sciences of the United States of America* 15: 773-9.

Zwicky, Fritz. 1929. "On the Red Shift of spectral lines through interstellar space." *Proceedings of the National Academy of Sciences of the United States of America* 15: 773-9.

Zwicky, Fritz. December 28, 1929. "On the Possibilities of a Gravitational Drag of Light." *Physical Review Letters* 34.

Extra reading: these are strongly recommended.

An interesting collection of information about the many open questions in physics is available at www.openquestions.com/oq-cosmo.htm.

Read it and you will be surprised at the degree of speculation and the admitted lack of understanding by many experts.

Methods of gravity determination using laboratory techniques and equipment are discussed at *http://mist.npl.washington.edu/eotwash/gconst.html.*

Evidence for the Big Bang—Remote Sensing Tutorial: http://rst.gsfc.nasa.gov/Sect20/A9.html.

Techniques for measuring distances, describes 26 methods: *http://www.astro.ucla.edu/~wright/distance.htm.*

Additional details are published and available in an earlier collection of essays/chapters on the subject in my first book. See Aisenberg, Sol, *The Misunderstood Universe*, 2009, 263 pages. A copy is available from your library or from Amazon.

You have permission to freely share past essays from the first book
(with proper attribution, and unchanged)
with your friends and colleagues,
and they are also encouraged to freely share these essays.

Appendix B

Curriculum Vitae of Sol Aisenberg, PhD

EDUCATION

Sol Aisenberg earned a PhD in physics from MIT with minors in physical electronics and electromagnetic theory. He graduated cum laude from Brooklyn College with a BS in physics and honors in physics after graduating from Brooklyn Technical High School. He also has held part-time appointments as a staff member at MIT in the physics department and in the Research Laboratory of Electronics (RLE). When he had time, he was also a part-time lecturer at Harvard Medical School and a visiting research professor in the bioengineering department of Boston University. He also held a first class radiotelephone operator's license.

He was elected as a member of the Phi Beta Kappa, Sigma Xi (science), and Pi Mu Epsilon (math) honor societies.

MANAGEMENT EXPERIENCE

After graduating from MIT, he joined the Raytheon Company and progressed to senior scientist at the Raytheon research division.

Aisenberg then joined Space Sciences Inc., where he progressed to become physics department manager.

After Space Sciences Inc. was acquired by the conglomerate Whittaker Corporation, he progressed to general manager, and then president and principal investigator of the space sciences division of the Whittaker Corporation (for over ten years).

After arranging the sale of the division to the conglomerate Gulf+Western, he was then president and principal investigator of the applied science labs division of Gulf+Western (for over nine years).

He then formed his own consulting company, the International Technology Group, with the Data Associates division.

TECHNOLOGY

Aisenberg is an applied physicist, scientist-inventor, and author. He is a generalist with experience in many fields including, but not limited to, plasma physics, electromagnetic theory, energy conversion, microwaves, electric arcs, instrumentation, lasers, optics, solid-state physics, radio frequency, and medical programs.

Aisenberg was a principal investigator for over five years in the artificial heart program of NIH, and also for many government contracts. He was also a co-principal investigator for over five years in a Joslyn Diabetes Foundation program for implantable glucose sensors.

Included are years of experience in obtaining, negotiating, and running programs for many government agencies including NASA, NASA Headquarters, NASA Ames, NIH, DARPA, and WPAF.

After forming his consulting company, the International Technology Group (ITG), he consulted for US and international companies. He obtained 510(k) new product approvals from the FDA for companies in the United States and in Sweden and England.

Aisenberg consulted for over a year as senior advisor on intellectual property for a major financial information company (Thompson Financial) and a global financial transaction company (Depository Trust Company). He also has consulted for law firms in Washington, DC, New York, and Boston on patent matters. As a result, Aisenberg is able to identify the technology that (a) can be patented, (b) that have market potential, (c) how to frequently bypass competitive patents, and (d) how to make his patents and client patents bulletproof.

He now is semiretired and independently inventing, patenting, and licensing inventions. One is to significantly increase the miles per gallon for internal combustion and diesel engines. He is also inventing a way to permit electric cars to drive unlimited distances with very fast refueling. He is also working on an invention to control cancer and drug-resistant infections without needing drugs, and to make it rapidly available and still comply with FDA regulations. In process are inventions to identify bank and store robbers and other persons of interest. He is also inventing a means to produce unlimited amounts of clean energy based upon his pioneer work starting in 1968-1972 for developing, demonstrating, and patenting thin films of carbon with the properties of diamond. He named them diamond-like carbon (DLC) and obtained four related US patents.

Aisenberg has over 143 publications, presentations, and reports, plus a number of invited presentations and over twenty-five US patents issued with others pending. He also received seven IR-100 awards for important new products and an award for research in helium-neon lasers.

He was a reviewer for a number of technical journals and has reviewed proposals for the National Institute of Health and for the National Science Foundation.

Aisenberg is a past member of board of directors of the Society of Professional Consultants. In the past, he served as a vice-chairman of the IEEE Boston Consultants' Network and also on the advisory group of the IEEE Boston Entrepreneurs' Network. He has presented invited lectures on diamond thin films, on inventing, on consulting, and on technology transfer.

Aisenberg is a current or former member of many professional societies including the American Physical Society, Division of Astrophysics, Division of Biological Physics, Division of Material Physics, Topical Group of Gravitation, the APS Vacuum Society, the Materials Research Society, the IEEE, Society of Photo-Optical Instrumentation Engineers (SPIE), and the Association for Research in Vision and Optics (ARVO), and the New York Academy of Sciences

He is a former member of technology transfer organizations, including the Licensing Executives Society (LES) and the Technology Transfer Society (TTS).

Aisenberg has been listed in editions of *American Men of Science*, *Who's Who in the World*, *Who's Who in Finance and Industry*, and *Who's Who in Technology Today*, among others.

Because of his interest in understanding the universe, he has spent some spare time in clarifying serious errors in the Standard Model of the universe. He then prepared and published a book to present his solutions as *The Misunderstood Universe* by Sol Aisenberg © 2009. The book is available from Amazon and in public libraries and should be of interest and use to future generations.

A new book by Dr. Aisenberg will show the effects of gravity in the cosmos (cosmic gravity), on its mysteries, and provide a new and truer model of the cosmos. This second book is in process and additional editions are planned.

Appendix C

Some fields of activity of Sol Aisenberg

This will demonstrate the ability to combine his knowledge and experience in many fields to contribute to the true understanding of the cosmos.

It is based upon experience in programs including federal contracts, foundation support, presentations at meetings, some publications, and issued patents.

Aisenberg has experience in many fields: plasma physics, electrical arcs, electrodes, electronics, measurements in plasmas, electromagnetic theory, microwaves and high power microwaves, defense programs optics, solar energy converters, magneto-hydrodynamic plasma jets, magneto-hydrodynamic energy conversion, gas lasers, biomaterials, medical instrumentation, artificial organs, blood-compatible materials, artificial hearts, diabetes control, implantable glucose sensors, and portable screening systems for lead poisoning in school children.

These include publications, reports, patents and presentations in fields such as but not limited to: Ultra-high vacuum systems, mercury arcs, Langmuir probe measurements in plasmas, paramagnetic resonance, microwave tests of spin Hamiltonians, failure of high-power microwave windows, photoemissive solar energy converters, cathode work functions, rare gas lasers, magneto-hydrodynamic plasma space propulsion, thin metallic films—deposition and diagnostics, theta-pinch lasers, high-voltage breakdown, cross-magnetic field accelerators, arc constriction processes, arcs in transverse magnetic fields, arc electrodes, electron emission processes at arc cathode spots, arc retrograde motion in crossed magnetic fields, dielectric properties of very thin insulating films, modification of re-entry plasmas, plasma boundary interactions, additives and plasma sheaths, absorptivity/emissivity measurement, semiconductors, software programs, deposition of single crystal and silicon nitride on silicon, plasma boundary physics, plasma accelerators, ion beam deposition of thin films of insulating carbon, diamond-like carbon (DLC), runaway electrons in a MHD energy generator, measurement of blood flow velocity, re-entry plasma diagnostics, implantable glucose fuel cell sensors, sensor for mass screening for lead poisoning, ion micro field emission from an arc cathode, magneto-hydro dynamically operated artificial heart pumps, alcohol breath sensors, fluorometric system for drug detection, artificial beta cell, implantable glucose sensor, ion beam deposited carbon coatings for bio-compatible materials, implantable oxygen sensor, approach for the treatment of diabetes mellitus, blood compatibility of ion beam-deposited diamond-like carbon (DLC), hermetic coatings for optical fibers, improved optical elements for high-powered lasers, moisture protection of strong optical fibers, novel materials for improved optical disk lifetimes, applications of the Meissner effect for superconductors, international technology transfer, locating and assessing available technology for transfer, how to determine the suitability for technology for commercial applications, measure and analysis of rapid eye movements, dyslexia, a simplified model of the universe, expanding gravity, plus dark matter and dark energy, and quasars.

Appendix D

Example: Diamond-like Carbon (DLC) (Pioneering work)

This will provide an example of Aisenberg's activities related to plasma technology and its application to diamond-like carbon (DLC) which he developed, demonstrated, and named in about 1968-1972. A list of related publications, reports, and presentations is provided. Also identified are four US patents related to DLC and issued to Aisenberg.

During the initial work on the insulating carbon films (DLC) in about 1968-1972, a plasma source to produce ionized carbon ions, argon ions, and sputtered carbon atoms was designed and constructed. It used a carbon anode and carbon cathode and was used to deposit the insulating carbon films These films were called diamond-like carbon (DLC) because these films had the properties of diamonds but were not bulk diamonds.

During the longtime periods of operation of the carbon ion source, we frequently had to stop and clean the exit of the source of carbon from deposits that clogged the exit. The deposits consisted of tiny carbon particles which may have been small buckyballs or carbon nanotubes. We did not have the equipment to study the structure of these particles, nor did we have the funds from government sponsors or from corporate to investigate this deposit. However, funds were obtained to study applications of the DLC, and some results are listed in the following.

Some of the Publications, Reports, Presentations, and Patents of Sol Aisenberg, PhD, which are Related to Plasma Technology and Diamond-Like Carbon (DLC)

This list is largely based on his experience and knowledge of plasmas, sputtering, and thin films.

Aisenberg, S. and V. Rohatgi. 1968. "Study of the Deposition of Single Crystal Silicon, Silicon Dioxide, and Silicon Nitride on Cold-Substrate Silicon." Interim report prepared for NASA Electronics Research Center under Contract No. NASA12-541.

Aisenberg, S. and R. Chabot. 1969. "Study of the Deposition of Single Crystal Silicon, Silicon Dioxide and Silicon Nitride on Cold-Substrate Silicon." Final report prepared for NASA Electronics Research Center under Contract No. NASA12-541.

Aisenberg, Sol. 1968. "The Deposition of Single Crystal Silicon Films on Cold Single Crystal Silicon Substrates." Proceedings of the 1968 Government Microcircuit Applications Conference (GOMAC), Gaithersburg, Maryland, October.

Aisenberg, Sol. 1968. "The Ion Beam Deposition of Single Crystal Silicon Films on Cold Single crystal Silicon Substrates." Proceedings of the Fifteenth National Vacuum Symposium, Pittsburgh, Pennsylvania, October.

Aisenberg, Sol. 1968. "Physical Processes For The Removal of Free Electrons in Re-Entry Type Plasma Sheaths." *Bulletin of the American Physical Society* 13: 1494.

Aisenberg, S. and R. W. Chabot. 1970. "An Ion Beam Deposition Technique for the Formation of Thin Films of Insulating Carbon." Presented at the Second Annual Symposium of American Vacuum Society and the New England Section of the Surface Division of the American Vacuum Society, Waltham, Massachusetts, May.

Aisenberg, S. and R. W. Chabot. "Ion Beam Deposition of Thin Films of Insulating Carbon." Proceedings of the Government Microcircuit Applications Conference (GOMAC), New Jersey, October 6-8.

Aisenberg, S. and R. Chabot. 1970. "Ion Beam Deposition of Thin Films of Diamond-Like Carbon." Presented at the Seventeenth National Vacuum Symposium, Washington, DC, October 20-23.

Aisenberg, S. and R. Chabot. 1970. "Ion Beam Deposition of Thin Films of Diamond-Like Carbon." Invited paper presented to New England Chapter, Thin Film Division, American Vacuum Society, December.

Aisenberg, S. and R. Chabot. January/February 1971. "Ion Beam Deposition of Thin Films of Diamond-Like Carbon." *Journal of Vacuum Science and Technology* 8: 1.

Aisenberg, S. and R. Chabot. 1971. "Ion Beam Deposition of Thin Films of Diamond-Like Carbon." *Journal of Applied Physics* 42: 2953.

Aisenberg, Sol. 1971. "Ion Beam Deposited Carbon Films." Invited seminar presented at Pennsylvania State University, Department of Material Sciences, May.

Aisenberg, S. and R. Chabot. 1971. "Deposition of Carbon Films with Diamond-Properties." Proceedings of the Tenth Biennial Conference on Carbon. Lehigh University, July.

Aisenberg, S. and R. W. Chabot. 1972. "Versatile Coating Systems for Ion Beam Deposition, Ion
Plating, Sputter Deposition and Sputter etching." Presented at the Nineteenth National Symposium of the American Vacuum Society, October. *Bulletin of the American Physical Society.*

Aisenberg, S. and R. W. Chabot. 1972. "Physics of Ion Plating and Ion Beam Deposition." Presented at the Nineteenth National Symposium of the American Vacuum Society, October. *Bulletin of the American Physical Society.*

Aisenberg, S. and R. W. Chabot. 1973. "Physics of Ion Plating and Ion Beam Deposition." *Journal of Vacuum Science and Technology* 10(1): 104.

Aisenberg, S. and R. W. Chabot. January 1973. "Investigation of the Properties of Thin Insulating Films Deposited with an Ion Beam System." Final Report No. AFCRL-TR-73-0176 for Air Force Cambridge Research Laboratories, under Contract Number, F19628-72-C-0291.

Aisenberg, S. and R. W. Chabot. December 1973. "Ion Beam Deposited Carbon Coatings for Bio-Compatible Materials." Final Report under Contract NIH-73-2919 for Division of Blood Diseases and Resources, National Heart and Lung Institute, NIH.

Aisenberg, S. and R. W. Chabot. November 1974. "Ion Beam Deposited Carbon Coatings for Biocompatible Materials." Comprehensive report prepared for National Heart and Lung Institute, NIH on Contract No. NIH-N01-HB-3-2919, Report No. Space Sciences Division-P-711CR.

Chabot, R. W. and S. Aisenberg. 1975. "Blood Compatibility of Ion Beam Deposited Carbon Coatings." Proceedings of the Twenty-Eighth Annual Conference on Engineering in Medicine and Biology.

Chabot, R. W. and S. Aisenberg. 1975. "Continued Measurement and Study of the Blood Compatibility of Ion Beam Deposited Carbon: Investigation of Dominant Physical and Chemical Factors." Prepared for Division of Heart and Vascular Diseases, NIH, Final Report.

Aisenberg, S. and M. Stein. 1977. "Diamond-Like Carbon Films—Factors Leading to Improved Biocompatibility." Proceedings of the Thirteenth Biennial Conference on Carbon, Irvine, California, July 18-22. *Extended Abstracts*, page 87.

Aisenberg, S. and M. Stein. 1977. "Continued Studies of Ion Beam Deposited Carbon as a Blood Compatible Material." Annual Report, Space Sciences Division-P-836A for Contract NIH-NO1-HB-3-2919 National Institutes of Health.

Aisenberg, S. and M. Stein. 1978. "Ion Beam Deposited Carbon Films and Factors Important to Improved Biocompatibility." Proceedings of the AAMI Thirteenth Annual Meeting, p. 6.

Aisenberg, S. and M. Stein. 1979. "Factors Important in Applications of Biocompatible Materials." Extended Abstracts of the Fourteenth Carbon Conference.

Stein, M. and S. Aisenberg. 1979. "Evaluation of Ion Deposited Carbon Films." Extended Abstracts of the Fourteenth Carbon Conference.

Stein, M., S. Aisenberg, and J. M. Stevens. 1980. "Ion Plasma Deposition of Hermetic Coatings for Optical Fibers." Proceedings of the Eighty-Second Annual Meeting of the American Ceramic Society, November.

Aisenberg, S. and M. Stein. 1980. "The Use of Ion-Beam Deposited Diamond-Like Carbon for Improved Optical Elements for High Powered Lasers." Proceedings of the Twelfth Annual Symposium on Optical Materials for High Power Lasers.

Aisenberg, S. and M. Stein. 1980. "The Moisture Protection of Strong Optical Fibers." Interim Report, RADC-TR-80-252, Rome Air Development Center.

Aisenberg, S., M. Stein, J. Stevens, and B. Bendow. 1981. "Ion Deposited Hermetic Coatings for Optical Fibers." Presented at the Fiber Optic Sensor Systems Workshop (FOSS), May.

Stein, M., S. Aisenberg, and J. Stevens. 1981. "Ion-Plasma Deposition of Carbon-Indium Hermetic Coatings for Optical Fibers." Proceedings of the 1981 Conference on Lasers and Electro-Optics (CLEO), June.

Aisenberg, S. and M. Stein. 1981. "Novel Materials for Improved Optical Disk Lifetimes." Proceedings of the Fifteenth International Technical Symposium of the Society of Photo and Electro-Optic Engineers (SPIE), August.

Stein, M. L., S. Aisenberg, B. Bendow, and BDM Corporation. 1981. "Studies of Diamond-Like Carbon Coatings for Protection of Optical Components." Proceedings of the Thirteenth Annual Symposium on Optical Materials for High Power Lasers, November.

Aisenberg, Sol. 1982. "Improved Hermetic Coatings for Optical Fibers." Radiation Curing VI Conference proceedings, September, Society of Manufacturing Engineers, ch. 12, pp. 14-32.

Aisenberg, Sol. 1983. "Properties and Applications of Diamond-Like Carbon Films." Proceedings of the American Vacuum Society, Thirtieth National Symposium, p. 132.

Aisenberg, S. and M. Stein. 1983. "The Moisture Protection of Optical Fibers." Final Report, Contract No. F19628-78-C-0180, RADC/ESM, Hanscom AFB. December.

Aisenberg, Sol. 1987. "Technology of Diamond like Carbon and its Applications." Invited lecture presented at a meeting of the American Association for Crystal Growth, Mid Atlantic Section, New Jersey, October 22.

Aisenberg, S. and F. M. Kimock. 1990. "Ion Beam and Ion-Assisted Deposition of Diamond-like Carbon Films." In *Preparation and Characterization of Amorphous Carbon Films*, Material Sciences Forum, vol. 52-53, edited by John J. Pouch and Samuel A. Alterovitz (NASA Cleveland, Ohio). Aedermannsdorf, Switzerland: Trans Tech Publications Ltd.

Aisenberg, Sol. 1990. "The Role of Ion Assisted Deposition in the Formation of Diamond-Like Films." *Journal of Vacuum Science and Technology A* 8(3): 2150-4.

Aisenberg, Sol. 1990 "Some Comments on Diamond-Like Carbon, Diamond, and Hard Carbon Materials Including Terminology, Deposition Processes, Composition, and Applications." Technical Note TN-90-01. *Applied Diamond Technology*.

Aisenberg, Sol. 1990. "Initial and Clarifying Publications Related to Diamond-Like and Hard Carbon." Technical Note TN-90-02. *Applied Diamond Technology*.

Aisenberg, Sol. 1990. "Diamond-like Carbon Deposition Technology for Improved Barrier Films." Invited paper, Fourth International Conference on Vacuum Web Coating, Reno, October 31-November 2. Bakish Materials Corporation, Englewood, New Jersey.

Aisenberg, Sol. 1991. "Practical Applications of Diamond thin films." Invited talk, New England Combined Chapter of the American Vacuum Society, March 20, Bedford, Massachusetts.

Aisenberg, Sol., A. Altshuler, and J. L. Sprague. 1991. "Physical Deposition of Diamond—Technology, Properties, and Applications." Presented at the Second International Symposium on Diamond Materials in the 179th Meeting of the Electrochemical Society, Washington, DC, May 5-10.

Altshuler, Anatoly and Sol Aisenberg. 1991. "Low Temperature Deposition of Artificial Diamond Material by Means of Halogen Chemical Transport Reaction." Presented at the 1991 Meeting of the Materials Research Society, Boston, Massachusetts. *Abstract Book*, p. 250.

Patents and Patent Disclosures Related to Diamond-like Carbon (DLC) with S. Aisenberg as Inventor or Co-inventor

Apparatus for Film Deposition, US Patent No. 3,904,505, September 9, 1975.

Film Deposition, US Patent No. 3,961,103, June 1, 1976.

Process for Coating Optical Fibers, US Patent No. 4,402,993, September 6, 1983.

Apparatus for Coating Optical Fibers, US Patent No. 4,530,750, July 23, 1985.

#

Appendix E

Related physical constants

1 parsec   pc = 3.262 light year

1 parsec   pc = 3.086567 × 10 exp 16 m

1 parsec corresponds to a parallax of one second of arc

Velocity of light   C = 2.998 × 10 exp 10 cm/sec

Velocity of light   C = 186,283 miles per sec

Astronomical unit   1 Au = 1.49597892 × 10 exp 10 13 cm

Gravitational constant   G= 6.6720 x 10 exp—8 dyn cm*cm/g*g

Index

www.ingramcontent.com/pod-product-compliance
Lightning Source LLC
Chambersburg PA
CBHW022003170526
45157CB00003B/1117

* 9 7 8 1 4 8 3 6 3 1 3 7 0 *